Lob Trees in the Wilderness

The Fesler-Lampert *Minnesota Heritage* Book Series

This series is published with the generous assistance of the John K. and Elsie Lampert Fesler Fund and David R. and Elizabeth P. Fesler. Its mission is to republish significant out-of-print books that contribute to our understanding and appreciation of Minnesota and the Upper Midwest.

The series features works by the following authors:

Clifford and Isabel Ahlgren

J. Arnold Bolz

Helen Hoover

Florence Page Jaques

Evan Jones

Meridel Le Sueur

George Byron Merrick

Grace Lee Nute

Sigurd F. Olson

Charles Edward Russell

Calvin Rutstrum

Robert Treuer

CLIFFORD AHLGREN
AND
ISABEL AHLGREN

Lob Trees in the Wilderness

The Human and Natural History of the Boundary Waters

University of Minnesota Press
MINNEAPOLIS • LONDON

Published by the University of Minnesota Press
111 Third Avenue South, Suite 290
Minneapolis, MN 55401-2520
http://www.upress.umn.edu

Printed in the United States of America on acid-free paper

Library of Congress Cataloging-in-Publication Data
Ahlgren, C. E. (Clifford Elmer), 1922–
Lob trees in the wilderness : the human and natural history of the boundary waters /
Clifford Ahlgren and Isabel Ahlgren.
p. cm. — (The Fesler-Lampert Minnesota heritage book series)
Includes bibliographical references.
ISBN 0-8166-3815-2 (pbk. : alk. paper)
1. Boundary Waters Canoe Area (Minn.) 2. Forest ecology—Minnesota—
Boundary Waters Canoe Area. 3. Forests and forestry—Minnesota—Boundary
Waters Canoe Area. 4. Nature—Effect of human beings on—Minnesota—
Boundary Waters Canoe Area.
I. Ahlgren, Isabel. II. Title. III. Series.
QH76.5.M6 A34 2001
333.75'09776'7—dc21
2001023492

18 17 16 15 14 13 12 11   10 9 8 7 6 5 4 3 2

God gives all men all earth to love,
     But since man's heart is small,
Ordains for each, one spot shall prove
     Beloved over all.
Each to his choice, but I rejoice,
     The lot has fallen to me—
In a fair ground . . .
                    . . . In a fair ground.

—Rudyard Kipling

To F. B. Hubachek, Sr., who has loved and championed a fair ground, this book is dedicated with deep gratitude.

Contents

.

Preface

Throughout the border lakes, in the days of the voyageurs, certain trees were singled out as "lob trees." Each grew on a prominent point or island, standing out from the distance and serving as a landmark to all who passed. Each was named in honor of a well-known explorer, fur company official, or voyageur who had performed a brave or noteworthy act. A nimble member of the crew clambered up the tree and lopped off branches from the central portion of the crown, leaving the middle bare with a tuft of branches above and below to make the tree more obvious. Lob trees were, in effect, wilderness highway markers, giving much needed orientation through the maze of lakes and streams, forests and muskeg. Most of the original lob trees were white pine, red pine, or occasionally white spruce. Most of them are gone now, but a few still can be found in the Quetico Provincial Park (see plate 1).

Representatives of nine native woody plant species serve as lob trees for the chapters of this book. We hope they will stand out as landmarks, symbolizing some of the human influences that have shaped the modern forest. Each species was chosen because of its ecological or sociological significance. Each is intricately involved in unfolding the history of human activity in the Boundary Waters Canoe Area (BWCA). Understanding past human influences is essential for sound, long-range wilderness preservation, protection, management, and utilization.

Since passage of the Wilderness Act in 1964, the BWCA has been part of the wilderness preservation system of the federal government. The area achieved full wilderness status in 1978. The act provided the following definition of wilderness:

A wilderness, in contrast with those areas where man and his works dominate the landscape, is hereby recognized as an area where the earth and its community of life are *untrammeled by man*, where man himself is a visitor who does not remain. An area of Wilderness is further defined to mean in this Act an area of undeveloped Federal land retaining its primeval character and influence, without permanent improvements or human habitation, which is protected and managed so as to preserve its natural condition and which (1) generally appears to have been affected primarily by the forces of nature, *with the imprint of man's work substantially unnoticeable*; (2) has outstanding opportunities for solitude or a primitive and unconfined type of recreation; (3) has at least five thousand acres of land or is of sufficient size as to make practicable its preservation and use in an unimpaired condition; and (4) may also contain ecological, geological, or other features of scientific, educational, scenic, or historical value. (P.L. 88-577, Sec. 2-C, U.S. Congress, 1964; emphasis added)

Webster defines "trammeling" as imposing limitations, restricting freedom. The following chapters retrace human footprints as they cross and recross the border lakes forests to determine whether evidence of their trammeling remains. Where such evidence exists, we mark it with a lob tree and consider implications for the destiny and potential of the BWCA. We will concentrate on the upland forest vegetation where human effects are thus far most evident, easily documented, and far-reaching in their implications. Evidence of human impact on birds and wildlife, aquatic habitat, watershed, and other aspects of the environment will be mentioned but not discussed in detail.

This book advocates no particular classification, status, or policy for the area. It describes human effects on the forest that must be considered in determining any status or administering any policy. It explores the background and reasons for forest changes and chronicles the concurrent development of human attitudes toward this unique portion of the northern forest. It champions no particular cause but that of sound ecological understanding of the relationship between human life and the wilderness forest.

Emphasis is placed on the BWCA—the United States portion of the border lakes country. However, where information is available and similarities or differences are noteworthy, the Quetico Provincial Park — the Canadian side of the border lakes wilderness—is also considered.

The notes gathered at the end of the book do not comprise a complete bibliography of the topics discussed. They include major publications and documentation of unique and significant findings that can serve as a starting point for further study. Common names of plants are used; scientific names will be found in the Appendix.

Acknowledgments

Many people have guided us to our lob trees, and without their help this book could not have materialized. They are not responsible for its shortcomings, however.

First of all, we are deeply grateful to F. B. Hubachek, Sr., for his support, encouragement, and inspiration in the cause of restorative wilderness research.

We especially thank the Trustees and Program Committee of the Wilderness Research Foundation who have so generously helped and advised us for over thirty years. Particular thanks to Frank H. Kaufert for his many years of support and guidance, to Henry Clepper for counsel and editing on this and many previous manuscripts, and to F. B. Hubachek, Jr., for his patience and faith in us during development of this book.

We are grateful to past and present personnel of the Superior National Forest, both the Duluth management staff and the ranger district field staff, for providing information and insights not previously available. Former supervisors of the Superior National Forest who were especially helpful were Harold Andersen, Craig Rupp, and John Wernham. Other past and present personnel who provided information and guidance are Joseph App, Sigurd Dolgaard, W. J. Emerson, Edward Hill, John Kernick, Russell MacDonald, Wayne Nicolls, Sheldon Norby, L. B. Ritter, Karl Sideritz, and J. W. White. Information was also obtained from John Anderson, Ralph Bonde, Ralph Graves, Darey Guard, Jake Licke, Sue Lawson, Carl Oyen, Stanley Smith, Eric Walberg, Myrna Waugh, Al Wolters, and Wilbur "Pappy" Wright.

Our appreciation to the University of Minnesota for use of facilities, advice, photographs, and assistance: John Carlson, George Freier, Alvin R. Hallgren, Henry L. Hansen, Donald Olson, John Ownby, Vilis Kurmis, L. C. Merriam, Raymond W. Darland, Richard Skok, and the staff of the College of Forestry, the University Photo Laboratory, and the Cloquet Forest Research Center.

Professional advice and assistance were also received from Robert Hagman, Charles Heiser, Arvo Kallio, Neil McKenna, the late Sigurd Olson, James C. "Buzz" Ryan, Robert St. Amant, and Shan Walshe.

Border lakes country residents who provided information or photographs include Lee Brownell, Thomas Chosa, Mr. and Mrs. Carl Gawboy, Ernestine Hill, Mr. and Mrs. William Trygg, Jr., and Florence Peterson.

Our sincere gratitude to those who reviewed and helped shape the early draft of the entire manuscript: Harold Andersen, Robert Cary, Jalmer Jokela, Wayne Nicolls, Craig Rupp, Milton Stenlund, and John Wernham.

Finally, special thanks to our two children for their childhood of slave labor in the cause of wilderness research, to Larry for photographic assistance beyond the call of duty and to Molly for painstaking review of all chapters.

Lob Trees in the Wilderness

Figure 1.1. Red pine lob tree. Photo by authors.

CHAPTER 1

Red Pine Lob Tree

INTRODUCTION

"Now *this* is really wilderness!" exclaimed the journalist as he stood beneath our lob tree red pine (fig. 1.1). One does sense an atmosphere of solitude and peace when standing beside it. Could this atmosphere be the essence of wilderness? It apparently was for the two California-based conservationists who stood with us that summer evening. One was a president's son, the other a well-known journalist; they had come to the area to gather information for an article on the canoe country for an outdoor magazine. They knew from experience in western wilderness what they wanted to see, and they saw it.

Wilderness? Could be. Our lob tree stands more than eight miles inside the official boundaries of Boundary Waters Canoe Area (BWCA) wilderness. But, along with its neighbors, it began life as a seedling in a Wisconsin nursery. The site on which it grows was logged in the early 1930s and cleared of brush. In 1936, the seedlings were nearly three years old when they were brought to northeastern Minnesota and planted here. Our lob tree, then, is still a youngster, and it is not a native. Already forty feet tall with a trunk eight inches in diameter, its size belies its youth and exotic origin. Together with neighboring trees planted at the same time, it is part of a deep pine forest. No other tree species are growing among these pines. A few shrubs and herbs provide a dappled lower layer, and the ground is covered with a deep carpet of needles.

Our red pine lob tree is younger and smaller than the original lob trees of the canoe country wilderness, and it stands away from the lakeshore, close to other trees, in a dense grove. The original lob trees selected by voyageurs during the eighteenth and nineteenth

Figure 1.2 Young red pine trees are typical of thin-soil, brush-free islands and shorelines. They make popular campsites that are easily damaged by overuse. Courtesy U.S. Forest Service.

centuries stood out singly along the lakeshore, on an island or on a point. Each one served as a landmark and was named in honor of a brave or prominent wilderness traveler. By lobbing or removing the center branches of the crown, the voyageurs made these trees easy to spot from a distance. The original lob trees helped guide the canoes along the uncharted maze of lakes and portages.

For over twenty-five years after planting, the forty-acre stand that includes our red pine lob tree was carefully tended and protected from competing vegetation. Lower branches were pruned; slow-growing, malformed, or crowded saplings were removed. Because the species is native to the border lake country, mature red pines are frequent in the area. Young stands like this one, however, are rare in nature. Unless planted, they are found primarily on rocky, burned-over islands or brush-free shorelines where the soil is shallow (fig. 1.2).

We chose a red pine lob tree for this first chapter because the species is one of the enigmas of this land, where the primitive and recent interact in a truly modern way. Like the land itself, red pine

has a sturdy appearance that seems to contradict its exacting requirements for survival. Both the natural reproduction of this species
and the wilderness quality of the land have been altered by human
activity. Survival of both in conditions resembling those of presettlement times depends on future human management and restraint.

Red Pine, a Demanding Species

Red pine has a natural range much smaller than those of the other
two pines of the boundary waters area. It is distributed naturally
between the 51st and 43d latitudes in a five-hundred-mile band from
Alberta through Minnesota to the Atlantic Ocean. In the Great Lakes
region, its northern limit is in the nearby Quetico; its southern limits
are in central Minnesota, Wisconsin, and Michigan. Jack pine, part of
the true northern or boreal forest, extends farther south into Indiana
and Illinois and north and west into Manitoba and Alberta. White
pine grows naturally as far south as Georgia.

Postfire ash on the soil surface inhibits red pine seed germination; germination of white and jack pine are not similarly inhibited.
Sod or litter from balsam fir and paper birch also inhibits red pine
germination. Seedlings do not thrive in shade and cannot compete
well on recently burned land where ash stimulates lush herb and
shrub growth. An abundant seed source, bare soil seedbed, and lack
of competition all must coincide in just the right way if natural red
pine stands are to develop. Since the removal of vast quantities of
seed trees during pine logging days, such coincidence is rare in the
border lakes country. There are occasional old survivors, scattered
patches of young red pine along shores and on islands, and the
gnarled wolf trees, rockbound in crevices where only winds belong.
These trees and the planted nursery stock often mistaken for natural
stands give red pine an important role in creating the atmosphere of
the canoe country "wilderness." However, the role masks human-
caused change in the behavior of red pine, as we shall see in later
chapters.

A Land of Controversy

Many aspects of the area have been subjected to constantly changing
influences ever since Europeans arrived, just as have the red pine.
These influences have interacted with the rapidly evolving human
demands and expectations for the area. One cannot fully compre-
hend modern BWCA ecology without appreciating these changes and

interactions. It is because of them that the canoe country has a penchant for producing controversy. No other wilderness has been the subject of so much disagreement, media attention, legislation, litigation, and regulation. Maintenance of the BWCA has been the focus of at least five special acts of congress, three presidential proclamations, two presidential executive orders, and two international treaties, as well as numerous directives and regulations by the Secretary of Agriculture and the U.S. Forest Service. Efforts to influence legislation governing the area have affected political careers, frustrated professional foresters, and stimulated intense lobbying by both industrial interests and preservationists from all over the country.

Modern controversy began about the turn of the century with efforts by General Christopher C. Andrews to dedicate some forested lands in northeastern Minnesota permanently to forestry. Although the state legislature, dominated by farm interests, would not restrict settlement of state-owned lands, Andrews' efforts resulted in the setting aside of 500,000 acres of Lake and Cook counties as federal forest reserve. When the Superior National Forest was created in 1909, this area was included in it. At that time, the Superior forest consisted of three separate units: a block south of Lake Saganaga, a narrow strip from Lac La Croix to the western edge of Basswood Lake, and a large central block, somewhat south of the other two. However, many choice portions of the present BWCA were privately owned and gradually acquired by the national forest in later years.

The very first studied attempts to protect the wilderness character of any national forest by limiting settlement and commercial recreational development began in the Superior National Forest. In 1922, Arthur H. Carhart, a Forest Service landscape architect, carefully evaluated the canoe country and recommended that all road building be halted in the remote northern portions of the forest. Many local residents, commercial developers, and conservationists disagreed. The long, heated battle had begun over what constituted the "greatest good for the greatest number in the long run,"[1] as directed in the original goals of the Forest Service. By 1926, however, road building had been restricted, and portions of the Superior National Forest were classified as "primitive areas" by the L-20 Regulations of 1929. In 1930, the Shipstead-Newton-Nolan Act banned power dams on the border lakes waterway and restricted timber harvest within two hundred feet of shorelines, roads, and trails. A similar act passed by the Minnesota Legislature in 1934 extended and strengthened this protection.

In 1939, the Forest Service U-Regulations created three more categories for undeveloped national forest lands: wilderness, wild lands, and roadless areas. Portions of the present BWCA were the only lands ever included in the roadless category. At that time, the canoe country was officially called the Superior Roadless Primitive Area. Still more congressional action—the Thye-Blatnik Act of 1948—authorized government acquisition of private and county lands within the area by condemnation. By then, the name had been changed to Superior Roadless Area. In 1949, a presidential order made the border lakes the only national forest in which air travel was banned to an altitude of no less than four thousand feet above the surface, except for emergency and patrol flights. In 1958, the name Boundary Waters Canoe Area became official.

In 1964, the Wilderness Act established a wilderness system within federal lands. The BWCA was the only area mentioned by name (twice) because of certain exceptions that pertained there: continuation of existing activities such as logging and outboard motor and snowmobile use, the need for construction and maintenance of small dams, and restriction of air traffic. Even with these exceptions, the BWCA was not named in the appendix to the act that designated wilderness areas, leaving it in bureaucratic limbo. Official wilderness status was not granted to the area until 1978. Inclusion of the BWCA among designated wilderness areas was controversial from the beginning. Even among wilderness advocates, there was disagreement about the appropriateness of such a classification for the BWCA.

Still another congressional act, the Fraser-Vento-Anderson Act of 1978, dealt specifically with the BWCA; it further increased restrictions on use and development, eliminating some of the exceptions given in the original Wilderness Act. Each of these legislative actions provoked heated controversy among local residents, industrial interests, conservationists, and preservationists. From 1948 to 1980, the courts were continuously involved with active or pending lawsuits contesting control of the area. Each of the acts changed the size, shape, and often the official name of the BWCA.

The constant controversy is directly related to the maverick quality of this wilderness. It is truly a land between—a land caught between east and west, north and south, concepts and contentions, future and past. It differs from many other wilderness areas in so many ways that generalizations, legislation, biological models, conclusions, and management plans tailored for other areas are like ill-fitting shoes when put upon this canoe country.

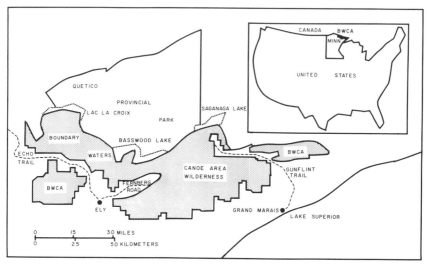

Figure 1.3. Location of Boundary Waters Canoe Area.

Not a Western Wilderness

For most people today, the word "wilderness" is associated with remote expanses of the western United States and of Alaska. The BWCA wilderness differs greatly from large wilderness areas of the West, which include well-consolidated blocks of land. Often these areas abut each other or adjoining primitive areas awaiting wilderness designation. Their interiors are far removed from many pressures of modern civilization. Many western wildernesses are located in large, sparsely populated states that buffer them from such pressures. In contrast, the BWCA is a long, narrow strip running about 110 miles east and west along Minnesota's border with Canada (fig. 1.3). This strip ranges from one to thirty miles in width and consists of three blocks separated by corridors along roads built long before wilderness designation—the Echo Trail or Ely-Buyck Road and the Gunflint Trail from Grand Marais on Lake Superior to Lake Saganaga. From time to time, other primitive roads have penetrated the area for logging and other access. With few exceptions, western wilderness areas have no access roads, and transportation is limited to foot or horseback.

Although somewhat buffered by the million-plus-acre Quetico Provincial Park to the north and the sparsely populated state and national forest land to the south, the BWCA is subject to industrial

Figure 1.4. Even for early travelers, the quiet beauty of wilderness lakes and the gentle land form made the border lakes country a less hostile environment than other wilderness areas. Courtesy U.S. Forest Service.

pollution and population pressures. It is within a day's land travel from the large metropolitan areas of Minneapolis, Milwaukee, and Chicago. Logging and mining towns fringe its southern border. Logging camps, resorts, and cabins have thrived for a time within that border.

Western wildernesses lack these influences. Most are lands of harsh geologic extremes—mountains, deserts, canyons. Early settlers moved through them as rapidly as possible to more fertile valleys and plains beyond. Northeastern Minnesota, however, with its many lakes and gently rolling, forested land, was a stern host but not a completely hostile one (fig. 1.4). The climate is harsh and the soils not agriculturally fertile; however, it reminded northern European immigrants of their homelands and provided a meager woodland existence. Some of them stayed. Their descendants found their livelihood in the iron mines, power plants, sawmills, and paper mills that developed nearby.

Figure 1.5. The border lakes from the air, a vista of lands and lakes. Names of these lakes reflect the appreciation and awe of early visitors. Courtesy U.S. Forest Service.

What's in a Name?

Names of landmarks reflect the attitudes of early travelers and settlers. In the West, names like Devil's Postpile, Hell's Canyon, Death Valley, and Desolation Ridge are common. In contrast, most border lakes proudly bear the names of the men who first found or used them, their ladies, Indian names, or descriptive terms like Blissful, Clearwater, Crooked, or Emerald. One finds the rare Devil's Elbow, Calamity, and Poverty lakes, and even Yabut Lake, indicating that the going did not always inspire poetry (fig. 1.5).

A Place to Have and Hold

The habitability of the area later complicated acquisition of land blocks suitable for wilderness designation and maintenance. Much land was already in private ownership when the Superior National

Forest was established. In fact, more than twenty miles along the Canadian border from Lake Saganaga to Basswood Lake were entirely in private ownership in 1928. This area is located in the portion of the BWCA most heavily used today. Maintenance of a primitive forest ecosystem could not be efficient in the original patchwork of private and federal holdings. The complicated, tedious task of land acquisition performed by the Forest Service within the limitations set by changing laws, available funds, and sensitive public opinion is often overlooked.

Because of the gentle land form and availability of woodland resources, the BWCA was subject to more human use than most large western wilderness areas prior to the preservation and recreation pressures of the 1960s and 1970s. Fur trade, the white pine industry, exploratory mining, and even commercial fishing and hunting all flourished for a time within this wilderness area. It now contains sizable elements of both uncut and cutover forest. Total acreage of virgin or uncut forest is often disputed. Preservationists claim that over half of the BWCA is covered with virgin forest; Forest Service employees working in the area before 1915 reported that most of it (except for old red and white pine stands) was cutover forest, logged between 1893 and 1910. Recent inventories place the limit somewhere between these two extremes.

Wilderness East

The location of the BWCA also makes it a land between—one lying between the relatively unaltered, massive, western wilderness areas and the smaller, more heavily used eastern ones. It was not until passage of the Eastern Wilderness Act in 1975 that significant federal lands east of the 110th meridian were incorporated in the wilderness preservation system—over a million acres, with provisions for later additions. Much of this land had been logged, settled, or used in other ways, but some wilderness qualities could still be protected and developed. The BWCA is located just east of the 110th meridian, but it and three smaller areas similarly located had already been included in the original Wilderness Act of 1964.

The BWCA is thus the oldest and by far the largest eastern area included in the federal wilderness preservation system. In 1976 it represented 45 percent of the total eastern wilderness acreage, but only 5 percent of the country's total. Most of the sixteen other eastern wilderness sites and the seventeen later to be included range from three to fifty thousand acres, with an average size of eighteen thousand acres. Only two sites exceed one hundred thousand acres, whereas the BWCA is now more than a million acres.

Northeastern Minnesota is also a land between eastern and western states in proportion of federal and private land ownership. Federal land available for wilderness designation is very limited in eastern states, increasing greatly from the Great Plains to the Rocky Mountains. Numbers change yearly. There were 14,443,584 acres nationwide included in the national wilderness preservation system in 1976.[2] However, the addition of Alaskan and other western lands dramatically increased this acreage by 1982 to eighty million acres, with other additions pending. In 1976, almost 90 percent of national wilderness preservation system land was under Forest Service juris-diction; 5 percent was administered by the Fish and Wildlife Service and 8 percent by the National Park Service. Recent additions have increased the proportion administered by the latter two agencies, with only about 33 percent of the wilderness areas now lying within national forests.

Changing Wilderness Concepts

Wilderness had a negative connotation from prehistoric times through early settlement and rapid expansion of the human habitat. At first, wilderness was viewed as a place of evil, fear, insecurity, and testing. Robbers and evil people symbolically hid in the wilderness; ethical people avoided it. The Bible and sacred writings of other major religions are replete with references to wilderness as the epitome of sin, evil, and darkness.[3]

As populations expanded, however, wilderness was seen as a source of livelihood. At first timidly, then with eagerness and greed, people began to tap wilderness resources. They viewed the riches of the wilderness as treasures placed on earth for human use. The Judeo-Christian mandate is given in Genesis 1:28, where on the sixth day of Creation, God tells Adam and Eve: "Be fruitful and multiply; and fill the earth and subdue it; and have dominion over the fish of the sea, over the birds of the air, and over every living thing that moves on the face of the earth." In the second Creation story in Genesis 2, however, God gives Adam and Eve responsibility for the care and protection of the Garden of Eden. Eden is a *garden,* however, suggesting the early Biblical writer's opinion that the truly desirable human habitat is a cultivated landscape. Similar ideas developed elsewhere, regardless of religion, philosophy, or state of enlightenment, as populations increased. Large portions of China, Mexico, Greece, and of other Mediterranean countries—especially those with semiarid or thin soils—have been cleared or overgrazed to the point of nearly complete destruction of natural forests and grasslands.

Wilderness concepts evolved much more rapidly in North America than they had in the Old World. Rapid population growth and technological advances played a key role in this evolution. When explorers arrived on the border lakes in the seventeenth century, most European settlers were huddled on the edge of a dark, forbidding wilderness along the East Coast. Nathaniel Morton, Keeper of Records at Plymouth Colony, noted in 1620 that the forests of Massachusetts were "a hideous and desolate wilderness full of wilde beastes and wilde men."[4] As late as the 1860s, Henry David Thoreau —considered part of the avant-garde of the wilderness movement— was not immune to these attitudes. He referred to wilderness as a place where "vast, Titanic inhuman Nature has got [man] at disadvantage, caught him alone, and pilfers him of some of his divine faculty."[5]

Even when pioneers penetrated the wilderness to hunt, cut wood, and clear land for farming, it was still viewed as a threat or a region to be conquered or eliminated. Much of the romance and adventure associated with the voyageurs of fur trade originates in the fact that these men ventured into such forbidding country.

As the frontier expanded westward in the 1800s, the eastern wilderness was rapidly subdued. Timber and minerals were extracted; lands were turned to agricultural use. Cities and towns burgeoned, and networks of roads and railroads spun out over areas once considered impenetrable. Even Thoreau was caught up in this view of wilderness as a source of mental, spiritual, and material resources for people. His famous statement "In wildness is the preservation of the world" is often quoted out of context. It continues: "Every tree sends forth its fibers in search of wilderness. The cities import it at any price, men plow and sail for it. From forests and wildernesses come tonics and barks which brace mankind."[6] It was just such a search for resources—timber and iron ore—that began the real expansion of settlement in northeastern Minnesota in the late 1800s. Agricultural development on the cold, thin soil was minimal. The wilderness shrank; cutover, nonagricultural land was everywhere.

Gradually, the nation approached its final frontiers and looked back on its path. Some people began to realize that the frontier had vanished; there was no more new land, and they must conserve what they had. Utilization of northern Minnesota's pine forests was at its peak when Theodore Roosevelt, Gifford Pinchot, and the conservation movement appeared on the woodland scene. But the nation also began to have time to play. The rigorous training required by the armed forces during World War I stimulated an interest in and need for vigorous outdoor recreation. People actually enjoyed being in the wilderness. Camping, boating, and fishing became popular, and the border lakes were well adapted to these uses. Wilderness became a playground, not a threat or a land to be exploited for its material resources.

The BWCA is thus also a land between ideas—a land caught between ever-changing human concepts of wilderness. These changing concepts have influenced human impact on the area. Early explorers passed quickly through the lakes, whereas later settlers learned to live within the wilderness, to use its resources and enjoy it. Finally, human impact changed to efforts to preserve and restore the wilderness.

Still another concept emerged as a stepchild of our overpopulation and of the tensions of modern, highly mechanized life: that of wilderness as a source of emotional and spiritual renewal. Some people came to see solitude as almost mystical and wilderness as holy. This intangible wilderness quality could be destroyed if too many people came to seek it. Wilderness had to be protected from overuse, preserved in a pristine state. Preservation became a crusade that was embraced with an emotional zeal new to the forests, mountains, and lakes of our nation. Although it originated in American wilderness experience, the pressure for wilderness preservation has now spread and become a worldwide movement.

Thus, the role of the relatively undisturbed forest was gradually reversed. Where thinkers and reflective people once gathered only in urban cultural learning centers, leaving the wilderness to the evil and primitive, now the solitude of uninhabited land was actively sought by many contemplative souls as a place for reflection, creative thought, and personal renewal. Evil and society's derelicts found anonymity in complex urban sprawl. Once more, the BWCA was caught as a land between—but this time between the pressures of resource harvest and the desire to enjoy the land recreationally, and between the impact of all human uses and the need to preserve its wilderness quality.

As more people came to the BWCA, many bringing different wilderness concepts, conflicts arose. But with the controversy, the increased knowledge and wilderness experience, many people expanded their concepts. We have often seen the individual development of wilderness awareness among college students of prairie and city origin entering the area for biological research and study at the summer field station of the Associated Colleges of the Midwest adjacent to the BWCA. At first the newcomer is fearful—afraid to leave the trail, afraid of the night sounds when the camp fire goes out, afraid of losing the way through the maze of channels and islands in the lakes. Gradually, as new and interesting wilderness products are discovered, the student becomes excited and very active, picking berries, fishing, gathering other natural foods, making ornaments, finding uses for wild things. Increased use and familiarity bring curiosity about plants and animals, their names and habits. Eventually, a close attachment to the area leads to a fervor for its conservation or preservation. Thus, in one short summer the ontogeny of BWCA users' wilderness concepts often recapitulates the phylogeny of the whole human approach to wilderness, evolved over thousands of years.

As these concepts evolved in northeastern Minnesota, some uncut forest remained that could be conserved and used differently. The lakes and forest lent themselves to increased recreational use. Potential for the solitude craved by wilderness enthusiasts remained.

Too Many People

Accessibility to large urban areas brought an influx of people. These human hordes seeking solitude and primitive recreation soon threatened the very wilderness quality they sought. Along major canoe routes on a July weekend, tents dotted every available campsite. Canoeists looking for a campsite faced a challenge similar to that of finding a motel room along the highway to Disneyland. Campsites eroded, beaches were defiled with human excreta, litter was strewn along portages, trees were needlessly cut, trees and rocks were carved upon and painted, and solitude disappeared. There developed a growing concern that the BWCA could be unwittingly loved to destruction.

The BWCA became the first national forest wilderness in which permits were required for entry. User permits provided more accurate records of recreational traffic, and eventually a quota system for heavily used routes was developed. Studies of user preference and impact on campsites were made. Cans and bottles were banned to reduce litter. A program of ecological awareness and education was developed for visitors. Canoeists obtaining their first permit at the

Voyageur Visitor Center in Ely stand sober and wide-eyed as they receive explanations about approved camping practices in a litany of instructions reminiscent of those given by an airline attendant at the beginning of a transoceanic flight. The BWCA has become a landmark in the development of programs to manage recreational use within wilderness areas. The changing status, regulations, and congressional acts involving the BWCA from 1926 through implementation of the wilderness acts from 1964 to the 1980s are testimonies to the unique combination of primitive character and heavy use. The former marks a contrast with eastern wilderness areas, and the latter sets this area apart from western ones.

A Highway of Lakes and Streams

The BWCA is also a land between the isolation of western mountain wildernesses and the easy accessibility of eastern areas. Absence of roads is one requisite for wilderness, regardless of how it is defined or where it is located. Indeed, the first real effort to preserve the remoteness of the canoe country was Arthur Carhart's recommendation in 1922 that road building be halted. For a time, logging roads were permitted in some parts of the area. Even without roads, however, nature provided an extensive water highway. Unlike wilderness areas accessible only by hiking, climbing, horseback riding, or a rapid dash by boat or raft on a river, the BWCA's network of lakes, rivers, and portages made human penetration and dispersal over the entire area relatively easy (fig. 1.6). Indian life centered on canoe travel. Voyageurs used these water highways for exploration and trade beginning in the seventeenth century. Steam-driven boats and motorboats later joined the canoes. The lakes provided a means of transporting logs to sawmills during the pine logging era. Travelers seldom move far inland from the water, but the network of more than eleven hundred lakes provides an extensive waterway that carries people over a wider range of the interior than in most other wilderness areas.

The lakes and streams also make the canoe country a land between in still another sense—a land caught between federal, state, and international jurisdiction. When the Webster-Ashburton Treaty of 1842 settled the long-disputed border between Canada and the United States, the boundary was drawn through many of the lakes and portages that form a direct and heavily traveled route from Grand Portage to Rainy Lake. The treaty established that the land

Figure 1.6. Part of the unique water transportation system of the border lakes. Courtesy U.S. Forest Service.

and water on the border be freely accessible to both Canadian and United States citizens; this right has been protected ever since.

For many years, state regulations governed all fishing and motorized use of the waterways that cover 17 percent of the area, although the most recent wilderness act now limits state control. The Shipstead-Newton-Nolan Act of 1930 protected those waterways from the development of extensive power dam systems. However, water levels and shorelines are influenced by the presence of more than seventy man-made dams, ranging from engineer-designed structures to simple rock and log accumulations. Many of these were created during logging days to help move pine logs. Because they served to shape the lakeshores and to form the land, they are now an integral part of the area.

When the glaciers retreated, they also left the border lakes area a land between—in this case between two watersheds. The Pigeon River watershed of the eastern portion drains into Lake Superior; the Rainy Lake watershed flows through most of the lakes into Hudson Bay via Rainy Lake and Lake of the Woods.

A Transitional Northern Forest Ecosystem

Even its original vegetation makes the BWCA a land between—a land between the boreal forests to the north and the northern hardwood and northern pine forests to the south. A few prairie and arctic plant species can also be found. Often characterized as a "tension zone" between zones, the land contains elements of both north and south that combine to confuse botanists and ecologists in their attempts to classify it.

More significantly, the constantly changing vegetation makes it a land between in yet another way—a land between stages of ecological succession. The area was constantly subjected to natural disturbances, primarily fire and wind, that temporarily destroyed portions of the forest. Telltale charcoal in soil and lake sediments, fire-charred stumps, and fire scars on old trees indicate that many portions burned every sixty to one hundred years. Long before Europeans arrived, the big pines rarely attained an age of three hundred years before being burned. Blowdowns were also common, because tree roots are shallow in this thin soil, and windstorms were frequent. So extensive was natural disturbance that a true climax or self-maintaining equilibrium forest has never been a significant part of the BWCA scene. In eastern forests, however, less flammable, deciduous species limit fires, thereby permitting some forests to develop to a climax. Although fire does influence forest succession in many western forests, trees that attain great age, such as the redwoods, often survive fires that burn smaller vegetation beneath them.

Human manipulation has helped make the BWCA a land between in one final way—a land between a natural forest and an intensively managed one. Extensive acreages were cut before the land became national forest. Until it achieved wilderness status, portions of the BWCA were subject to traditional management practices, including supervised cutting and planting. Red pine, jack pine, and white spruce were often planted following logging. The effects of logging and tree planting are thus interwoven into the development of the modern wilderness ecosystem.

The presence at one time of extensive private holdings and resorts also contributed to the area's uniqueness in the wilderness system. Gradually phased out with passage of the Thye-Blatnik Act of 1948, the privately owned developments also influenced the forest's ecology. Resorts are now gone; many of the resort sites have been planted with trees, and the casual traveler can see no evidence of settlement. Ghosts live on, however, in introduced plant species that, like those species introduced during horse logging days, have become a small but interesting part of both the history and vegetation of the canoe country.

The BWCA exists, then, as a maverick among wilderness areas of the United States in size, location, land form, accessibility, ecology, human attitudes, use, and related problems. One former supervisor of the Superior National Forest described these problems:

> Since the BWCA was not pure wilderness, it did not permit "normal" management prescriptions as applied to other wilderness areas in the system. I personally know of no other wilderness areas in the system which had so many exclusions, laws, agreements, and regulations applied to them. All wilderness areas have some degree of complexity and conflict. However, to my knowledge, none match the BWCA in complexity and controversy.[7]

It is our hope that by examining these complexities from a biological viewpoint, we can contribute to a better understanding of the area. Such an understanding must develop among both the public and those who manage the BWCA. As Gifford Pinchot, first chief of the Forest Service, said in 1907: "National forests exist today because the people want them. To make them accomplish the most good, the people themselves must make it clear how they want them run."[8]

Most people agree that we need national forest wilderness and that it needs our concern. Besides concern, however, management of the various wilderness areas requires much insight. Each wilderness is a distinct unit; generalizations among them can be misleading. In succeeding chapters, we will explore the forests of the BWCA within the unique context of its various plant species and their individual abilities to adapt to a changing environment. We will also consider the interaction of human activity and environment as it affects the forest. We hope to provide a new approach for understanding how the area has changed, how it continues to change, and how it can be maintained.

Figure 2.1. Ground juniper lob tree. Although the sacred juniper described in this chapter is gone now, the illustration shows the typical growth habit on sunny, exposed rocks where this species normally grows. Photo by authors.

CHAPTER 2

Sacred Juniper Lob Tree

FLORA

To visit our sacred juniper ring, we would have to go back in time at least twenty years and beach our canoe at the northeast end of Washington Island in Basswood Lake. A high, pine-covered point guards the entrance to a wind-sheltered waterway between the island and the mainland. The shaded, soft forest floor beneath the pines is bare except for a carpet of needles. Ground juniper (*Juniperus communis* var. *depressa*) grows in a distinct circle about ten feet in diameter, the low, bristle-needled branches radiating outward from the circle's arc (fig. 2.1). Nearby, some small, cedar-shingled shelters mark Ojibway grave sites (fig. 2.2).

    The lob tree is gone now. We last saw remnants of it in 1954. Now, during most of the summer, a tent covers the actual site where our lob tree once grew. In recent years, vandalism and recreational use have destroyed the Indian grave shelters and juniper ring. Their disappearance was inevitable, perhaps, because Washington Island has been a focal point over the years for traders, loggers, and campers. If one looks closely, signs of Indian and fur trade activity, logging, scientific studies, silvicultural manipulation, and heavy recreational use can all be found.

    According to legend, the juniper ring once played an important role in secret rituals of the Midiwiwin, the powerful Ojibway medicine society that gathered here. Efforts to uncover the exact meaning of the ring proved futile. Today's canoe country Indians say they've "been told" that their ancestors planted the ring. We only know that the ring once grew on this shady forest site where the species would not normally have appeared unless planted. We also know that the

Figure 2.2. Grave of Ojibway chief covered with birch bark, near Ely in the early 1900s. Note circle of plants pushed into bare ground surrounding the grave. Although this ring is probably hazel, not ground juniper, the use of the ring of branches is believed related to establishment of sacred juniper ring on Washington Island. Courtesy Lee Brownell.

planting of some kind of evergreen at or near gravesites was an Ojibway custom. Folktales say that the legendary Nanabojou planted a cedar on the grave of his grandmother, Nokomis.

Although it was a low shrub and not a tree, we have chosen the juniper ring as lob tree to symbolize human influence on the flora because it is one of the earliest known human manipulations of plants in the BWCA and because it evokes the irretrievable past. This Ojibway planting, however, had little influence on forest vegetation.

Basswood Lake Blueberry Ridges

Across the protected water channel and three miles north, a rocky ridge runs northeast along United States Point. The vista from this high, open ridge is spectacular, revealing the vastness of the Quetico stretched out to the north (fig. 2.3). Scattered clumps of stunted jack pine and a few aspen, birch, and spruce make up a sparse forest cover, with heavier forest along the drainage lines. On rocky outcrops,

Figure 2.3. United States Point, Basswood Lake, site of Ojibway burning for blueberry culture. Photo by authors.

aromatic sweet fern, warmed by the sun, makes the air fragrant. In many places, the ground is covered with blueberry bushes. In the 1800s and possibly earlier, this ridge was periodically burned over by the Ojibway to keep the forest back and maintain the open conditions in which blueberries thrive. Dried blueberries were an important part of the Indian winter diet. During years of berry crop failures, illnesses with symptoms now recognized as typical of certain nutritional deficiencies were a grimly accepted fact.

Indian control of the vegetation left more permanent and obvious changes here than did the planting of the ground juniper ring on Washington Island. In many places the thin soil has been burned and washed or blown away, leaving only irregular root and moss mats on rocky outcrops. Charcoal remnants of past fires can be found in the depressions. Natural fires also occurred, of course, but those set by Indians were more frequent here than elsewhere.

Remnants of Resorts

About ten miles south of the blueberry ridges, another open, rocky point warms in the morning sun. June canoeists and fishermen

Figure 2.4. Lilac growing on abandoned resort site twenty years after buildings were removed. Photo by authors.

stopping on the point for a lunch-and-stretch break are surprised to find a sturdy lilac blooming luxuriantly where rock meets the pine forest (fig. 2.4). This bush was planted here, but, unlike the sacred juniper, it is not native to northeastern Minnesota. It was brought here in the 1930s to grace the dooryard of the main lodge at Wenstrom's fishing camp. The cabin and other buildings were removed in the 1960s but the lilac remains, a witness to yet another chapter of human history in the border lakes country. Although the plant has a few sprouts, it has not spread and does not seem to be affecting the flora significantly.

Farther down the lake in a sheltered cove on the south shore of Hoist Bay is another abandoned resort site (fig. 2.5). The buildings were also removed in the 1960s. There are no lilacs here. The forest appears to be closing in to such an extent that a casual observer noted: "The canoeist paddling Basswood Lake cannot discern the shores where the Peterson Lodge entertained fifty guests at a time in 1955. But the loons know things have changed; they have returned to Hoist Bay."[1]

Now, if that casual observer had been an old-timer, he would have known that the loons never really left Hoist Bay (fig. 2.6). Nesting on the unpopulated shores across the bay, they regularly watched the evening return of fishing boats to Peterson's in the

Figure 2.5. Peterson's Resort, Basswood Lake, in the 1930s. Courtesy Florence Peterson.

Figure 2.6. The loon, longtime border lakes resident. Courtesy U.S. Forest Service.

Figure. 2 7. Goutweed, a naturalized, cultivated escape growing on Peterson's Resort site twenty-five years after abandonment. Photo by authors.

1940s and 1950s, as they had earlier watched birchbark canoes and now aluminum craft. Had the casual observer been a little more observant, he would have spotted some remnants of resort days, strikingly obvious from a passing canoe. A dense, light green and shining white carpet stretches back from the former boat landing, covering the old resort grounds and extending into the forest beyond (fig. 2.7). This green and white carpet is made up of thousands of plants of goutweed, locally called snow-on-the-mountain. Like the lilac on Wenstrom's Point, goutweed is not native; it was planted by the Petersons in the 1930s to bring a touch of home to their remote lodge. Unlike the lilac, this aggressive species has become naturalized and has been spreading vigorously for more than twenty-five years since the lodge closed in 1955.

Very few herbs or shrubs grow in association with goutweed because it produces substances toxic (allelopathic) to many native species that would otherwise be found here. Alder, planted red pine, and older trees grow above it, but these were well established before the goutweed began to spread. The plant grows well in sun or shade, and its carpet now extends for six hundred feed along the shore and four hundred feet back into the woods behind the lodge site,

Figure 2.8. Note extensive invasion of goutweed and absence of other ground vegetation. Photo by authors.

eliminating native plants as it spreads (fig. 2.8). A taller, aggressive form with no color variegation on the leaves has appeared in shadier spots. It is spreading even more vigorously, giving good indications that it is a truly naturalized part of the flora. This shade tolerance is unique among naturalized plants, most of which are characteristic of sunny habitats.

Wilderness Immigrants, a Hardy Group

Sacred juniper, blueberry ridges, lilac, and goutweed are small examples of the human role in shaping the flora of the BWCA. They are not isolated examples: many others can be found on other lakes and portages, each with an interesting story of plant adaptability and human initiative. The plants just described were manipulated or planted deliberately, as were lily-of-the-valley (fig. 2.9), peony, rhubarb, chives, cultivated iris (plate 4), lupine, sweet William, forget-me-not, Tartarian honeysuckle (fig. 2.10), and other domesticated plants still surviving in the BWCA with varying degrees of success. A host of others arrived more surreptitiously with horses and oxen, railroads and sleds, moccasins and boots. Others are still

Figure 2.9. Lily-of-the-valley on Skidway resort site, Basswood Lake, fifteen years after it was abandoned. Photo by authors.

Figure 2.10. Tartarian honeysuckle planted at resort site on Wenstrom's Point, Basswood Lake, twenty years after abandonment. Photo by authors.

arriving on boots and sneakers, in packsacks, tackle boxes, and lunch bags, or carried by birds, animals, water, and wind. A weed now common along heavily used portages, the plantain, was called "white man's foot" by Indians who noticed that it seemed to appear wherever the newcomers walked.

A careful inventory of BWCA flora reveals that ninety-two species, or over 11 percent of the 817 known flowering plant species, have been introduced or naturalized from outside the area. These invaders are a varied group, representing twenty-three plant families and seventy genera. Grasses make up the largest number, seventeen species. Many of them, including timothy, brome grass, redtop, and fescue, were associated with horse and oxen feed used by loggers and settlers in the early 1900s. The legume family is well represented by a number of vigorous clovers also introduced with horses. Yellow hop clover, for example, can be found naturalized and spreading aggressively in the open, sunny expanses of some blueberry fields. Sixteen species of the composite family have been introduced. These are mostly camp followers or weeds common in open places. Among them are dandelion, thistles, oxeye daisy, yarrow, yellow dock, sheep sorrel, and burdock. Things have changed since Warren Upham reported in 1884 that oxeye daisy, Canada thistle, and brome grass were just beginning to appear in southeastern Minnesota, introduced from the east.[2]

Many camp followers produce windblown seed. However, they seldom advance beyond the sunny, open trails and campsites because they require light for good growth and are also sensitive to organic acids in the forest floor litter.[3] These recent arrivals increase and become established where blowdown or fire exposes mineral soil and lets sunlight in. Plant immigration is a continuing process. In the course of our thirty years in the area, we have watched the introduction and spread of such species as goat's beard, gypsyweed, dandelion, and orange hawkweed.

Most introduced species came from Europe. Only eight of the invading plants are natives of other sections of the North American continent. Some of these might eventually have appeared in the BWCA even without human intervention. Two of them, however—wild plum and Virginia creeper—are escapes from cultivation. They are the only nonnative woody plants to become naturalized. The lilac and a few planted exotic trees cannot be considered naturalized because they have not yet spread beyond the original planting sites.

See How They Come

A variety of dispersal mechanisms influence the distribution of plant species in the wilderness. One striking example is the distribution of the native path-follower rush. Any moccasin, boot, hoof, or paw entering the area has been whipped by this tiny, resilient, grasslike plant on some game path or trail. Its seeds are borne in tiny capsules at the end of whiplike stalks. The harder these capsules are slapped by passing feet, the farther the seed is spread. On dewy mornings or rainy days, the capsules absorb moisture and float the seeds in gluey balls that adhere to passing feet and legs, carrying the seed some distance before they come to rest in soil. The sturdy little rush does a good job of holding soil along the edge of trails and portages. As other growth closes in when the trail is no longer used, the path-follower rush declines and is spread to other paths.

Some other native plants have more modest dispersal methods and correspondingly limited rates of spread. Bishop's cap, for example, is an inconspicuous little species producing several delicate flowers with lacy petals on a single stalk arising from a perennial, evergreen basal leaf rosette. Although the plant is found in many conifer forests, it produces flowers and fruits best in aspen woods. The fruit capsule opens into a shield shaped like a bishop's miter on which are borne six to ten shiny, black seeds. In a sudden summer shower, the bishop's cap acts as a splash cup or resounding board for raindrops that are deflected and splash off, carrying seeds a short distance from the parent plant.

Dispersal of some species remains a mystery. Poison ivy, assiduously avoided by most wilderness travelers, produces an inconspicuous, dry, smooth fruit not likely to be picked by human passersby, although it is used as food by birds, deer, bear, and small mammals. Somehow, this species finds its way most often to sandy, sunny, heavily used portages and campsites. It is less frequent in similar sandy, sunny spots away from human traffic. Perhaps camp-following birds and mammals are responsible for this selective dispersal.

Still another intriguing mystery is the cloudberry, a yellow-fruited, arctic raspberry sometimes found growing in sphagnum along the margins of dense lowland black spruce stands (fig. 2.11). Circumpolar in arctic realms, it covers vast acres of tundra in northern Scandinavia; it is also found on a few sites across central sections of Canada and in Maine and New Hampshire. The plant has not been reported from the Quetico. Its southernmost collection site in the interior of this continent is a black spruce swamp near Basswood

Figure 2.11. Cloudberry, a yellow raspberry typical of tundra. Distribution in BWCA is limited, source is unknown. Photo by authors.

Lake, more than 100 miles from the nearest colonies in the Thunder Bay District of Ontario. One can only speculate about the bird, mammal, or glacial finger that planted cloudberry in the BWCA. There is the remote possibility that this species, like the rare, large form of lingonberry, was introduced to the southern portion of its range by Finnish and Swedish settlers who prized it in their homeland.

Some Come, Some Go

The recent history of BWCA flora is not limited to the introduction and spread of species. Some species are disappearing. Even our best intentions occasionally threaten the existence of certain plants. Shore plantain, a small, inconspicuous inhabitant of gravelly or muddy shorelines, was referred to by the Harvard taxonomist M. L. Fernald as one of the rarest plants in North America. Until 1957, it was known in the BWCA only from a collection made in 1886 by the Cornell botanist Liberty Hyde Bailey. By 1957, the local botanist Olga Lakela had discovered six colonies growing in the area. Quetico Park Naturalist Shan Walshe has recently reported it to be frequent in the Quetico.[4]

A thriving shore plantain colony was observed and protected from 1957 to 1964 by the Wilderness Research Center staff in a nearby bay on Basswood Lake. In the early 1960s, however, local efforts to improve the mallard population resulted in the annual spring introduction of hundreds of brooder-hatched mallard ducklings to the bays surrounding the tiny shore plantain colony. The grain-fed ducks thrived, migrated in the fall, and many returned to nest nearby in subsequent years. More were introduced each spring for about ten years. The infertile bays provided meager food for large populations of breeding ducks. The mallards soon decimated some of the plant species of the fragile shoreline habitat. Within five years, shore plantain disappeared from the bay. The fate of this obscure species on one small bay on Basswood Lake is not world shaking, but it illustrates what happens when the attempt to manage one facet of nature does not include consideration of the effects on other parts of the ecosystem. Further effects of the mallard population included increased organic matter in the shallow acquatic environment; altered macrofauna, microfauna, and aquatic plant life; and accelerated eutrophication. The bay has been deserted by humans and brooder-hatched mallards for almost ten years now, but the aquatic environment still bears subtle but long-term changes as souvenirs of those "ducky" days.

Other infrequent species found on border lake shorelines could be threatened by changes in water level and quality, competing vegetation, and fauna. Upland ecological niches have been threatened by such practices as promiscuous burning, indiscriminate logging, and intentional or accidental introduction of exotic plants and wildlife. The thin, sandy loam soils of the area lack variety, and geologists agree that the edaphic (soil) variations of the BWCA are not sufficient to determine plant distribution except where rare clay veins and lowland bogs are concerned. Most natural habitat variations are thus caused by differences in moisture, light, and associated species, all factors that can be altered by human practices.

Our knowledge of diminishing, threatened, or endangered species in the BWCA is fragmented. Most of the attention given to endangered species is directed at animals. Morley's list of endangered Minnesota plant species includes more than fifty that occur in the Superior National Forest.[5] In 1979, the Minnesota Natural Heritage Program considered thirty-nine species as either endangered, threatened, sensitive, or rare. Both lists, however, contain some prairie or boreal plants that are frequent in other places. In 1975, the Minnesota Department of Natural Resources found no BWCA plant species to be endangered or threatened. The absence of good floral distribution and frequency studies, or of records of the prelogging flora for comparison, prevent an accurate evaluation of the status of BWCA species. Evaluation must be based on the frequency with which a plant has been found. Plant collecting, like travel, is often concentrated along waterways, slighting the more remote interior and giving no picture of total species distribution. In addition, species like Abbey's hybrid rock fern are naturally infrequent or rare, but they may be thriving where they occur and in no danger of extinction. Others that are more frequently encountered and therefore not considered endangered may be declining at a rapid rate as their habitats are altered.

The reduction of habitat variety by human use and management is a threat to the preservation of the natural flora. We can acquire the techniques needed to manage and maintain wild lands in the future, but we must acknowledge and understand the continuing effects of past human activities and the potential long-range effects of manipulation on the total flora. Subtle interrelationships may be overlooked until changes are irreversible; other links in the ecosystem are still unknown.

The Flowering Season

The distribution of plant species varies not only with habitat preference and seed dispersal mechanism but also with the season in which seed is produced. In the north country, flowering begins in late April or early May with the opening of alder, willow, hazel, and aspen catkins. Rising maximum temperatures trigger these early flowerings. Buds begin to swell when midday temperatures rise above 68° F for several consecutive days, even if nightly frosts occur.[6] Dates of first spring warm-up vary by several weeks from year to year, varying in turn the onset of flowering.

The catkin-bearing species are followed by the spring ephemerals, mostly low, flowering herbs of white, pink, or purple. Among these are the wood anemone, spring beauty, hepatica, goldthread, and violet that bloom in late May and early June. The heath-type shrubs also bloom at this time. Flowering is stimulated by sunlight reaching the forest floor before competing vegetation leafs out to shade it and by the long photoperiod of early spring days. Although these species may blossom slightly later if spring is unseasonably cold, their flowering dates are more consistent than those of the earlier flowering, temperature-dependent species. Many spring ephemerals disappear from the woods soon after flowering and exist as bulbs, roots, corms, or seeds in the forest soil until the next spring when they again sprout, bloom, and fruit. Many produce seeds that drop to the ground or are spread by insects and small mammals in their search for food. The ephemerals are well adapted to the northern climate and forest. All are native; the spring life-style does not lend itself to migration.

Species flowering from late June through early July are more varied and brightly colored. They include taller species—the buttercups, columbine, clintonia, bunchberry, fireweed, the orchids, most grasses. Although some of these plants are not native to the area, they are familiar to the observant canoeist. Many are pollinated by insects, but self-pollinated and wind-pollinated plants are also common. Most species blooming in midsummer are stimulated to flower by the long-day photoperiods of early summer when the sun is above the horizon for more than sixteen hours. Their flowering dates do not vary much from year to year unless a hard June frost nips the first buds. Seed dispersal methods include everything from the production of edible fruit that is dispersed by animals to the snapping slings of wild geranium and the splash cups of bishop's cap and some mints. Many seeds and fruits form burs and stick-tights that adhere to passing animals. Some are disseminated by wind.

In late July and early August, there is a temporary decrease in number of species in flower. The late flowers seen from mid-August until frost are mostly tall, sun-loving species with blossoms in muted shades of yellow, purple, and white. Many of these species exist in the vegetative state in the forest, flowering only along trails or in openings where sunlight reaches them; the big-leaved aster along canoe portages is a notable example. Many produce windborne seeds and fruits that persist on the plant well after snowfall, providing winter food for birds and small mammals. The fleshy fruits of several species, including mountain ash and highbush cranberry, also provide winter food. If left unharvested, these fruits are available as food when snow melts in early spring. Occasionally the fleshier fruits ferment, adding intoxication to the frenetic activity of birds and some small mammals during the mating season. Many introduced species are among these late-flowering plants.

Toward a BWCA Floral Inventory

The long process of identifying and recording the BWCA flora began with early explorers, some of whom collected a few specimens for study or made notes of what they saw. Their contributions to botanical knowledge were fragmentary, however, for most of their attention was directed at general features of land, animals, and Indians. Early botanists in Minnesota tended to stay close to the Mississippi River. The border lakes country was ignored botanically until the Geological and Natural History Survey was extended north in 1878 and 1879 and the first actual lists of plant species recorded in northeastern Minnesota were published.[7] In 1886, Liberty Hyde Bailey collected throughout the border lakes country and kept accurate records, but the first exhaustive collections were begun by Olga Lakela about 1940.[8] These collections have been continued and expanded by us at the Wilderness Research Center, which now has the most complete BWCA herbarium.[9] Duplicate specimens of many of the center's collections have been sent to herbariums at the Universities of Toronto, Minnesota-Twin Cities, and Minnesota-Duluth. The flora of neighboring Quetico Provincial Park in Canada is also well documented in collections made by Claude Garton of Lakehead University and by Shan Walshe.[10]

One of the discouragements of present-day plant collecting in the canoe country is finding that most new discoveries are recently introduced weeds that are common farther south in settled areas. Most native species were noted before 1965, but continuing collections add to our knowledge about the changing flora.

A Comparison of Floras

About 1800 flowering plant species can be found in Minnesota. Their distribution, however, is not uniform throughout the state. According to University of Minnesota checklists, the largest number of species occurs in the unglaciated or driftless portions of southeastern Minnesota—the fertile and moist areas of prairie and big woods near the Iowa border in Houston and Winona counties, where more than 960 species have been found. The least variety occurs in the western, true prairie counties such as Pipestone and Kittson, where fewer than 700 species have been reported.[11] The number of species is also lower farther north, with fewer than 700 species listed for the Quetico.

More than 800 species have been reported for the BWCA thus far. Interesting comparisons can be made between the floras of the BWCA and the Quetico to the north. Some 230 taxonomic entities of flowering plants (or 27 percent of the flora) of the BWCA have not yet been found in the Quetico; however, the Quetico has only 61 entities (or 10 percent of its flowering plant species) not represented in the BWCA. Of the species unique to each area, about 15 percent are introduced plants. Most plants unique to the Quetico are rare species or plants of special habitats. The majority of those unique to the BWCA are plants from southern or settled areas. The BWCA has a more uniform soil type than the adjoining Quetico, encompasses more transition from northern pine and mixed hardwood forests to the true boreal forest, has more token remnants of southern associations, and has more pressure of introduced plants from populated areas to the south.

Trees are less likely to be overlooked by plant collectors than are some obscure forbs and grasses, and they help shape habitat niches for smaller species. Numbers of tree species are thus relatively accurate, providing interesting confirmation of our comparisons of botanical variety in different parts of the region:

	Number of Tree Species
Pipestone County (southwestern prairie)	9
Kittson County (northwestern prairie)	16
Clay County (west central prairie)	19
Quetico Provincial Park, Ontario (boreal)	24
BWCA (northern pine and hardwood, boreal)	25
Clearwater County (northwestern forest)	30
Blue Earth County (south central, prairie and big woods, deciduous)	32
Houston County (prairie and hardwood, driftless)	45
Winona County (east central prairie and hardwood)	48

The flora in the BWCA is larger than in the prairie or boreal forest. It is smaller than in the transitional area between deciduous woods and prairie in southern Minnesota. However, the flora of even the most remote area is never static, always changing. Nature has provided mechanisms for a slow but steady change in distribution of plant species. The increased mobility and the industrial, agricultural, and population pressures that accompanied European migration have brought other mechanisms that speed up distribution of some species, destroy others, and in other ways may be changing the flora of the wilderness.

Figure 3.1. Jack pine lob tree. Photo by authors.

CHAPTER 3

Jack Pine Lob Tree

FOREST FIRE

To reach our jack pine lob tree, one must clamber up a rocky ridge overlooking a small lake. The climb is worth the effort; a vista of the surrounding lakes and forest is spread before the rugged outcrop. The hot sun bakes this exposed site on a summer day, and our lob tree offers only sparse shade (fig. 3.1). It is a small tree, no more than thirty feet tall, somewhat bent by winter winds. Its needles are coarse and short, scattered in pairs along the branches. Almost every branch bears several pairs of hard, gnarled cones; the oldest ones farthest back from the tip are grey and weather-beaten after ten to twenty years of exposure to the elements. Yet many of these old cones tenaciously protect viable seeds, releasing them only if fire opens the hard, resinous cone scales (fig. 3.2).

The ability of jack pine to become established on exposed sites, to produce cones when only a few years old, and to release seeds most efficiently after fire make it the obvious choice as lob tree for a discussion of fire in the border lakes country. No other tree species is so well adapted to survive on land swept repeatedly by wildfire. Although scattered jack pine seedlings occasionally develop in sunny, unburned openings, all real jack pine forests in the BWCA are of fire origin. Jack pine, in pure stands or associated with aspen, birch, black spruce, and balsam fir, occupies more than one-third of the entire BWCA, a striking testimony to the close association of forest fire with the vegetation of the border lakes country. Fire has been such an integral part of BWCA forest succession that jack pine, BWCA, and wildfire are closely linked in the minds of many plant ecologists.

Figure 3.2. Jack pine cones when closed (A), as they appear on the tree, and opened (B) after exposure to heat. The inset shows seeds released from opened cones. Photos by authors. Originally published in T. T. Kozlowski and C. E. Ahlgren, *Fire and ecosystems* (New York: Academic Press, 1974).

On rocky sites with thin soil, jack pines like our lob tree remain stunted, crooked, and low. On deeper soil, dense postfire stands grow straight and tall, often reaching heights of seventy-five to eighty feet and surviving one hundred years or more before stand decadence and windfall become problems. Rejected by early pine loggers and, in some parts of the BWCA, saved from modern pulpwood harvest by cutting restrictions, jack pine is the most common pine species in the remote portions of the BWCA.

Jack pine has been a much maligned tree. Early surveyors tried to ignore it. Early foresters considered it a weed. Its wood is coarse grained and weak, much less valuable for lumber than is red or white pine. J. A. Fitzwater, supervisor of the Superior National Forest in 1910, told of a ranger's ditty that went, "There, there, little jack pine, don't you sigh. You'll be a white pine by and by."[1] Even as late as 1955, Minnesota botanist C. O. Rosendahl slurred our lob tree species by calling it: "the least valuable of our pines. As it has little beauty to commend it, it is seldom cultivated."[2] Yet our lob tree and the rocky outcrop to which it clings typify the canoe country to many visitors.

Fire has played a major role in border lakes forest development from ancient times to the present. Knowledge of fire is therefore fundamental for understanding other forest influences. For this reason, we will discuss it before considering the gradual evolution of the wilderness forest and the other factors that have helped shape it. This chapter explores the role of fire in limiting and shaping the resources available for human use. It also considers the human role in altering the causes, frequency, size, and shape of fires, seasons of burning, and paths of postfire forest succession.

Both human activity and wildfire can alter forest cover abruptly; it is thus inevitable that these two sources of change should interact and that some of these interactions should create new patterns in the ever-changing forest mosaic. A comparison of fire trends during the recent past (1930-1980) with evidence of presettlement fire reveals fascinating differences that have important, long-range implications for the management of fire in this wilderness ecosystem.

Evidence of Presettlement Fire

A pattern of repeated fires emerged in the border lakes country as soon as flammable postglacial vegetation developed. This pattern continued for thousands of years, according to evidence from charcoal particles found in layered lake sediments (varves). Measurements of charcoal in varves obtained from one lake in the eastern portion of the BWCA establish an interval of sixty to seventy years between major fires in the forest surrounding the lake, with a range of twenty to one hundred years.[3] Fire frequency apparently varied with rainfall and temperature. Pollen samples obtained from other lakes and from peat bog layers reveal that jack pine was prevalent during periods when intervals between fires were short; white pine increased when the intervals were longer.[4] Natural fires still burn dry, open jack pine sites similar to those probably present in early postglacial times.

Fire history since the end of the seventeenth century is enhanced by data obtained from stumps, surviving stands of old red and white pine, scattered old trees, and from charcoal in the soil. Such evidence was obtained in the BWCA by M. L. Heinselman[5] and in the Quetico by G. T. Woods and R. J. Day.[6] The approximate time since the last major fire is determined from the age of most of the older trees in a stand. However, Woods and Day found a ten year age span within old postfire red pine stands supposedly of uniform age. In addition, red pine does not germinate on fresh ash during the first few years after a fire. It sprouts several years later, when rain and melting snow have leached the surface ash layer. For these reasons, fire dates determined by stand age are only approximate.

Trunks of some old trees bear "catfaces" or fire scars. The age of wood at the base of the scars, as determined by counting tree rings, aids in dating fires. A fire scar is often formed when, with ample ground fuel, a running surface fire encircles a tree. The wind and air currents created by the fire cause an eddy of air on the lee side of the trunk where fire can linger. Increased fire intensity on this side produces enough heat to scorch and kill vital trunk tissue. On the windward side, however, fire passes quickly, the trunk remains undamaged, and the tree survives. As the tree continues to grow, new tissue forms along the margins of the fire-damaged portions of the trunk, and a characteristic catface is formed (fig. 3.3). To initiate such a scar, a fire must move fast and not linger, or the entire trunk would be killed; temperatures above 60° C coagulate the protein in the living cells of a red pine trunk.

Trees bearing fire scars may be found along the edges of sites burned during major past fires. Scars may also be found on trees in areas burned over by light, ecologically insignificant fires, or where finger like projections of surface fires once radiated out from larger, intense fires. Therefore, not all scarred trees represent sites of major fires, and to interpret them all as such can be ecologically misleading.

Figure 3.3. Catface or fire scar on red pine trunk, formed by heat damage to cambium and bark as ground fire moved past tree. Courtesy Vilis Kurmis.

Major fires remove all vegetation and stimulate types of postfire succession discussed in a later chapter. However, surface or ground fires do not kill the overstory. They temporarily reduce or eliminate aboveground parts of understory species, but they stimulate sprouting of aspen, birch, and brush species. Light burning spreads these species so that when a major fire does occur that kills the overstory pines, the understory, broadleaf species are more vigorous and highly competitive, drastically reducing the possibility of tall pines seeding in.

Some foresters question the accuracy of fire dates determined by fire scars on trunks because the age of surviving, unburned annual rings varies with fire intensity and the rings in newly forming portions of the trunk adjoining the fire scar are often irregular and interspersed with corky callous tissue.[7] For the year 1871, for example, Heinselman reports no fire scars and no major fires within the million-acre BWCA. However, this was a year of extreme drought during which the very large Manistee and Thumb Lake, Michigan, fires; the tragic Peshtigo, Wisconsin, fire; and even the famous urban Chicago fire all occurred. In addition, a prairie fire one hundred miles long burned from Breckenridge to the Big Woods in southern Minnesota, and a U.S. Signal Corps meteorologist noted in his log that there were "unparalleled fires in northeastern Minnesota."[8]

Because fire dates obtained from this evidence are only approximate, Canadian workers prudently compiled early Quetico fire data by decades, not specific years. We will also interpret Heinselman's BWCA specific fire dates by decades. The BWCA and Quetico fire history information is otherwise similar in most important respects.

Ancient Fires and Early Human Life

Heinselman determined that 83 percent of the old forests originated between 1681 and 1894. He related their origins to specific fires within that period. Many of these fires occurred during early European exploration and fur trade (1731-1833 for the BWCA, 1699-1875 for the Quetico), suggesting that fires increased in frequency with early human activity. Grace Lee Nute believed that increased fires in the late nineteenth century were in part responsible for the disappearance of caribou because of the destruction of reindeer moss (lichen), the main food of this species.[9] However, pollen and charcoal studies provide evidence that fires were actually less frequent during the past five hundred years, indicating that very early European influence on fire incidence was probably minimal.

Indians used fire to drive game, control insects, and maintain blueberry fields, but their use of fire in this area before the late 1800s is unknown. Very light use of the area by Sioux before the Ojibway and low Ojibway population since then make this influence negligible, if Indian life-style was similar to that described in the late nineteenth century. Had Indians used fire more extensively, this use would be reflected in their lore, their continuing customs, and in explorer observations. We may therefore assume that most fires prior to settlement were natural and caused by lightning.

A comparison of the pattern of presettlement fires with that of modern fires reveals the extent of human influence on fire as a major determinant of forest succession. In both the BWCA and the Quetico, fires increased in frequency and size during the early settlement period (1860 to 1910, and to 1920 for the Quetico) during the fur trade, settlement, heavy human westward migration along the Dawson Portage in the 1870s, mining, and logging.

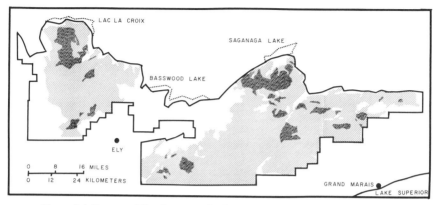

Figure 3.4. Pattern of fire frequency in the BWCA. White portions were unburned by any major fires between 1610 and 1880. Light-shaded areas burned one to two times during that period. Dark-shaded areas burned three or more times.

Nonrandom Distribution of Early Fires

When Heinselman's detailed history maps of early fires[10] are super-imposed on each other, they reveal a nonrandom distribution of fire before the logging era (fig. 3.4). Major fires occurred repeatedly in the western portion of the canoe country, near and south of Lac La Croix and in the eastern section, west and south of Seagull Lake. In the central portion—the entire area from Basswood Lake south—however, no evidence of major fires between 1610 and 1972 was charted. The Canadian fire history study also reveals a nonrandom distribution of early fires in the Quetico, with repeated fires north and east of Lac La Croix, a section of southeastern Quetico, and few or no fires in the central portion north of Basswood Lake.[11]

There may be several reasons for the lack of major fires in the central portion of the BWCA and in the Quetico just north of it. First, since the Minnesota portion was heavily logged in the early 1900s, many old stands were cut and evidence of fire was removed. Some small, old red and white pine stands remain, however, and should provide clues. Furthermore, evidence of fire is lacking in the adjacent portion of the Quetico where extensive logging did not take place at that time. In our experience, fire scars on old trees are infrequent in this portion of the canoe country. Lightning strikes are common on tall, exposed pines, but most strikes only split the affected tree. A few ignite and smolder in the trunk, but they rarely ignite surrounding vegetation.

Second, loggers were attracted to the north central portion by the quality of the tall pine stands there. Photographs of white pine harvested in the Winton area and crumbling stumps still remaining indicate that much of the forest consisted of large pines. Their growth implies that the land was free from fire for at least two hundred years.

Finally, Marschner's map of the original vegetation shows the presettlement forest mosaic of this portion to be largely white and red pine.[12] In contrast, the western portion of the BWCA south of Lac La Croix and the eastern portion southwest of Seagull Lake contained some pine but had more substantial acreages of jack pine, aspen, and birch, all characteristic of lands burned more frequently. The chance of fire in this more open vegetation is much higher than in dense, tall pine stands where the microclimate is more humid, fuels more moist, and where there is less solar drying and air movement.

The basic difference in type of presettlement forest in the east, west, and north central canoe country could also be related to differences in site and soil qualities. However, BWCA soil variations are not large and do not greatly influence tree species distribution, although they do alter growth rates, size, and stand quality. Distribution of canoe country tree species is more closely associated with length of time between disturbances, especially fires, than it is with soil type.

The fire rotation period or average interval between reburnings of an area equivalent to the million acres of the BWCA is often estimated at sixty to one hundred years. However, substantial acreages in the central portion did not burn for more than three hundred years, while portions elsewhere burned several times in that interval. This nonrandom, patchy distribution of fire invalidates any generalized fire rotation figure based on an average for the entire area.

Dry Lightning—A Key to the Fire Pattern

In modern times, certain parts of the area have a history of frequent fires and associated short rotation forest types; they are also centers of lightning-caused fires. W. J. Emerson, fire control officer in the Superior National Forest during the late 1940s and 1950s and later in Forest Service Region 9, which includes twenty states from Maine to Minnesota and south to West Virginia and Missouri, reported that during his years working with fires in national forests, the Superior—

especially the BWCA—was known to have a unique problem with dry lightning fires.

Dry lightning occurs when, because of high clouds and associated dry atmospheric conditions, rain either does not form or most of it evaporates before reaching the ground. Most dry lightning fires in the canoe country occur in July and August, seldom in the early spring and practically never in the fall. In contrast, "wet" lightning accompanied or followed by significant precipitation can occur throughout the year, even during winter snowstorms, but primarily between May and October. Wet lightning rarely ignites surrounding forest fuels, although it may scorch a tree or snag.

Fire maps of the past thirty years indicate two portions of the forest with pronounced, repeated patterns of lightning-caused fires. One is around Seagull Lake, incorporating two or three townships on the south and west sides. The other includes about two townships south of Lac La Croix. Substantial lightning fires spread out from these two centers. Throughout the rest of the Superior National Forest outside the BWCA, dry lightning fires are much less frequent.

During the past twenty years, 96 percent of all July lightning fires and 74 percent of all August lightning fires in the Superior occurred within the BWCA. However, only about 60 percent of the Superior's total fires occurred in the wilderness area. The dry lightning problem is rarely encountered in other forests in the Lake States, although it is common in western mountain forests, where it causes the majority of remote fires,[13] and in portions of Manitoba. Certain south central Florida forests have June dry lightning fires, but they are not a problem elsewhere in the eastern half of the United States.

The two areas with most frequent, repeated fires during presettlement times (fig. 3.4) radiated out from the two centers of modern dry lightning fires, suggesting a similarity between presettlement and modern lightning fire distribution. Ecologically important also is the prevalence of jack pine and aspen-birch types in these frequently burned areas of high lightning incidence. Still another minor center of increased presettlement fires is evident south and east of the Kawishiwi River; the postfire jack pine that developed there became the basis for the Tomahawk and other pulpwood sales discussed in chapter 7.

In contrast, fires of human origin during the past thirty years are scattered throughout the border lakes country, including the central portion where in the past lightning fires were scarce and the interval between fires was longer than in the eastern and western portions. Many recreation-caused fires (campers, fishermen, hunters) are clustered near access points—lower Saganaga and Basswood lakes, Lakes 1, 2, and 3, etc. Modern fires are also scattered throughout the Quetico, with concentrations where recreational use has been heavy — Basswood Lake, Lac La Croix, and Crooked Lake. Human activity dispersed fire more evenly than did lightning, the natural agent of fire. As a result, patches of recent postfire vegetation are more evenly scattered over the area.

No one knows exactly why fires caused by dry lightning are more frequent in certain portions of the border lakes country than in other portions. Superior National Forest veterans tell us that the prominent zones for lightning fire in the eastern and western portions usually have abundant bare rock and open, light vegetation that dries out quickly and is easily ignited. Old-timers speculate that dry lightning could be more frequent where concentrations of certain minerals, especially iron and copper, are found.

Physicists have found that igneous rock protruding or covered with thin, dry soil attracts dry lightning.[14] Experienced canoeists avoid camping on rocky, exposed points when there is the possibility of a thunderstorm. The presence of iron or copper ore in the rock does not seem to increase the lightning attraction. Dry lightning does not occur on similar, mineral-laden sites on the Upper Peninsula of Michigan. However, the Michigan area receives considerably more rainfall and as a result has somewhat different vegetation.

Lightning fires are more frequent during drought years when low-pressure centers move rapidly over dry ground surface and vegetation. Frequent electric charges are emitted, little rain falls, and the possibility of ignition is high. Although lightning strikes both trees and snags, it also hits bare ground, igniting duff and sparse, dry, low vegetation.[15] The resulting fire can then creep to areas with more fuel. Most work on the actual ignition of lightning fires has been done in the northern Rocky Mountains, where less than half of the lightning fires originate from strikes on trees; the rest start on exposed ground. Some strikes may ignite and smolder in snags or duff for days or weeks until fuel dryness, wind, and low humidity permit the fire to spread.

Most lightning strikes do not start forest fire. Many strike trees that crack or splinter but do not burn. Others ignite, smolder, and go out because of rainfall, fuel moisture, or lack of wind. In the central portion of the BWCA, every summer we have found old pines and white cedars struck by lightning, cracked and splintered. Some have charred centers but little or no burned area around them. Indians tell of watching for lightning-struck cedar to provide them with splintered-out beating sticks for wild rice harvest.

Some lightning strikes are "cold", meaning that they will not start fires even in dry fuel. Cold lightning results from very short peaks of electric discharge, whereas "hot" or igniting discharges produce an initial peak followed by about one hundred milliseconds of continuing current.[16] No one has yet determined whether a difference in conductivity of rocks and soil in portions of the BWCA prone to dry lightning increases incidence of hot lightning. However, since dry lightning is limited to certain portions of the border lakes country and open rock with light vegetation is common throughout the area, something besides rock and thin vegetation must cause this unique phenomenon.

Human Influence on Fire Season

Most natural fires in the BWCA occur in July and August, when the most dry lightning also occurs. Occasionally, fires may smolder in duff or snag until September before flaring up and expanding. Some remote fires in May or late fall are attributed to lightning. However, experienced fire officers believe that these are often fires for which causes cannot be determined. Some fires of obscure origin in May, for example, have turned out to be "string jobs" — fires set by beaver trappers along ridge lines to spite rival trappers and draw game wardens to their illegal trapping.[17]

If the season of dry lightning was the same in presettlement times as it is now, we can assume that most presettlement fires also occurred in July and August. Fires in spring and fall, then, are primarily caused by human activity. The Little Sioux Fire of 1971 is a recent, striking example of a spring fire, burning more than fourteen thousand acres and starting during slash burning on a logging site. The fall fire season in September and October has also produced memorable fires of human origin elsewhere, including the Peshtigo, Wisconsin, and the Cloquet and Hinkley, Minnesota, fires, which began as forest fires, and even the strictly urban Chicago fire. During the tall pine logging era beginning in the late 1800s, state regulations required loggers to burn slash. This, and the associated general increase in human activity, caused fires in spring and fall.

From 1948 to 1978, two-thirds of all fires on the Superior National Forest were of human origin. Most of these occurred, as do lightning fires, in July and August, thereby intensifying the summer fire season. However, many human-caused fires occurred in the

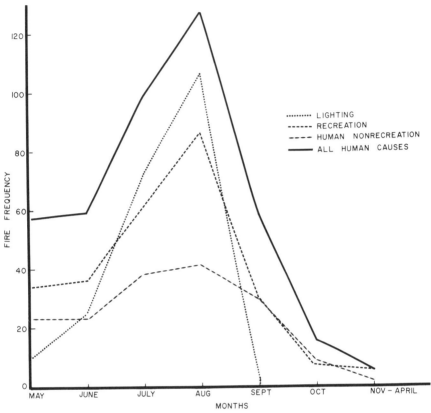

Figure 3.5. Monthly distribution of fires of various origins in the BWCA, 1948-1978. Data taken from Superior National Forest fire records.

spring and fall, extending the total fire season several months (fig. 3.5).

In the border lakes country, spring fires typically cause more damage than do fall fires, primarily because the long spring days provide more hours of sunlight when higher temperatures, wind, and low humidity favor fire spread. In contrast, during the short fall days, the midday peak of fire danger is much shorter. Most of the increase in human-caused fires occurs in the spring after the snow has melted, when the forest floor is dry, new vegetation has not yet emerged and the days are long, providing extended periods of sunlight and accompanying high fire danger. Fires at this time of year stimulate sprouting of aspen, birch, hazel, and alder more than do late summer and autumn fires.[18] In addition, spring fires destroy developing conifer conelets and buds, often eliminating the next conifer seed crop. In this way, the increased spring fires since settlement may have favored an increase in deciduous trees and shrubs over coniferous species and therefore could have been responsible for the increase of aspen in the total forest pattern or mosaic. Aspen is now present and increasing in more than 70 percent of the forest in both the BWCA and the Quetico.

Drought and Fire Incidence

Human activity has also extended the fire incidence more evenly over the years than during presettlement times. Modern lightning-caused fires in July and August occur most frequently in drought years, probably because the fuel is dry and dry lightning storms without rain are most frequent during those years. According to the pollen and charcoal studies, presettlement fires were also more frequent during dry periods. In contrast, human-caused fires since 1950 often occurred during years with average annual or growing season rainfall.[19] Therefore, drought years are not necessarily the only years with frequent fires in modern times. This occurrence of fires during non-drought as well as drought years has produced a greater variety of postfire successional stages, adding one more human effect to the forest mosaic. This greater successional variety interacts with the smaller fire size since fire fighting began. The result is a smaller, more varied age class pattern in the wilderness forest mosaic that could be of value in reducing threat of future forest damage from insects, disease, blowdown, and even fire.

Fire Weather

An understanding of weather patterns is critical in predicting fires and interpreting their behavior. Fires can start in the border lakes country during minidroughts or short dry periods accompanied by high pressure and wind conditions that are conducive to fire. According to W. J. Emerson, classic fire weather occurs most often when a Hudson Bay High approaches and forest fuels are dry. The high is accompanied by southwesterly winds, high temperatures, and low humidity, driving the fire northeastward:

> The presence of dry fuel creates intense combustion, resulting in
> thermal convection and a "fire storm," with the strong, dry wind
> adding oxygen to the process. Embers of burning material are blown
> ahead of the rapidly moving fire front causing advance "spot" fires that

further accelerate the fire's northeasterly advance. This advance results in the southwest to northeasterly shaped burn so typical of the border lakes. Sometimes the fire fans out or broadens as the high passes, the wind switches and drives the fire southeastward.[20]

Heinselman related the elongated fire shape, slanting to the northeast, to the northeasterly orientation of swamps and water bodies acting as fire barriers.[21] However, this fire shape is also found in portions of the area with no lakes as well as in places where fires jump lakes by advance spotting. Therefore, fire experts believe that water barriers have little influence on fire shape when evaluated against the massive effects of the Hudson Bay High, dry fuel, and associated weather conditions.[22]

Fire Suppression

Fire suppression was introduced into a portion of the border lakes country when the Superior National Forest was established in 1909, and it began in the Quetico ten years later. Modern fire fighting attempts to reduce the damage of both natural and human-caused fires. Fire control was the major activity of the national forest organization in the early days. It was rough work, and in drought years the crews worked against almost insurmountable odds.

The successful control of fires in remote areas is a tribute to the stamina, courage, and woods wisdom of the early fire control personnel. Few details of their activities have been published, but tales of old-timers on the Superior rival those of voyageurs for sheer adventure.

Despite slow travel by foot or paddle, lack of airplane surveillance, and poor communications, the early fire crews on both sides of the border significantly reduced the acreage that was burned annually. Heinselman reported that during the presettlement period before fire control, 1.5 percent of the virgin (uncut) forest burned annually, whereas from 1910 to 1972 only 0.05 percent burned annually. Woods and Day estimate that before fire suppression the acreage equivalent of the entire Quetico Provincial Park burned every 66 years, but that with the rates of suppression achieved from 1920 to 1976, this acreage would have burned every 700 to 1,667 years.

From humble beginnings with a few fire towers (fig. 3.6) and with telephone lines, canoes, shovels, axes, and buckets, fire fighting methods have advanced to involve airplanes (fig. 3.7), helicopters, bulldozers, water bombing (fig. 3.8), and specialists in meteorology, fuels, and fire behavior. After World War II, increased use of airplanes and later of helicopters greatly increased the speed and

Figure 3.6. Mullberg Fire Lookout Tower, 1916. Courtesy Lee Brownell.

Figure 3.7. U.S. Forest Service Beaver on BWCA fire patrol. Courtesy U.S. Forest Service.

Figure 3. 8. Forest Service airplane using water bomb to extinguish forest fire. Courtesy U.S. Forest Service.

efficiency of fire fighting. Costs mushroomed with increased technology, from an average of $136 per fire in the 1950s to $185 in the 1960s and to more than $2,000 per fire in the 1970s. The total amount spent for fire control in the Superior National Forest rose from $150,000 for the entire twenty years from 1949 to 1969 to $3 million for the decade of the 1970s. These figures also include funds for fire prevention.

In the BWCA, as elsewhere in the national forest system, fire prevention has been the most effective means of controlling fires of human origin. The early rangers on the Superior gradually developed techniques by education, law enforcement, and fire hazard reduction aimed at stopping fires before they start. Throughout the national forests, Smokey Bear and other fire prevention programs are credited with saving more money and reducing more of the area burned than any other factor. Within the BWCA, Forest Service workers, guides, and outfitters have cooperated in educating the public on camp fire safety and adherence to fire-ban rules during periods of high fire danger.

Modern fire size in the BWCA is determined in part by land form and the nature of recreational use. Many BWCA fires originate close to water. Typically, they are camp fires left unextinguished, often on islands or points with natural water barriers. These lunch and campsite locations are often sheltered from wind and midday heat. They are quickly reached by seaplane, boat, or canoe, and a well-equipped crew of two or three workers can extinguish them before they get out of control. As a result, the Superior National Forest has a record of a higher percentage of small fires (less than one-quarter of an acre) than any other eastern national forest.

Modern fire fighting begins with a sound knowledge of fuel conditions, weather influences, and the shape, direction, and rate of fire spread. Working with this knowledge, fire control officers seek to contain and limit the acreage burned. Success depends on efficiency of the crew, fuel dryness and quantity, the vagaries of wind and weather, and a little luck. A high hazard condition lacking water barriers was graphically described by W. J. Emerson:

> It is true that *no one*, with any kind of equipment now available, gets in front of an intense, fast-running fire-storm type fire and stops it. We *do*, however, do alot to limit and hold the fire to a size much smaller that it would ultimately reach were we not to attack it.
>
> . . . . The fire probably takes off in a dry southwest wind in dry fuels. It usually makes its run at about the time we are arriving at its source, where the plane or tower that discovered it has sent us. Often we cannot see the fire front because of smoke, and we don't waste time trying to get in front of it. Typically, however, the two flanks or sides are not burning as hot, so we start building line with bulldozers, hand tools, pumpers, etc., from the rear or fire source, along or near both flanks, progressing toward the head which usually drops and slows down at night when winds drop, humidity rises, and temperatures fall.
>
> Our objective (and it's critical) is to hold this fire so that it does not take off the next morning and make a new run during the severe daytime weather conditions. Often, in the BWCA, the rough, rocky terrain, heavy brush, and absence of access routes make a full night fire fighting operation nearly impossible, so we hold crews on the rear and flanks to improve and burn out the line, hopefully to have all edges under at least partial control when daylight arrives. We organize all night and move crews and equipment to the more or less temporarily tame head of the fire at the crack of dawn before the wind rises. They work to build control lines wide and clean as close to the slow-moving front as they can work. If we are successful in building a good control line around the fire head, burning out all material between the line and fire edge, we have a good chance, depending on midday wind and humidity, of containing the fire and holding it to that size. It won't be

easy, as spot fires usually start across our line. In the BWCA, our airplane observers usually will locate small spot fires burning quite a distance downwind, across our fire lines.

Now, *if* we get the classic 90 degree wind shift to the northwest after the front goes by, then our entire southeast flank will be our fire front (down wind edge). So we'd better have the southeast flank in good shape! If not, our fire may turn out to be a round one, or a square one, instead of nice and slim and long.

This is not just theory—it works that way, or it does if we have people in charge who know how to make it work. Well-trained crews and plenty of the right equipment are essential. I recall fires in the BWCA and elsewhere in the Superior . . . where we did succeed in doing this. I also know of fires where we did not, and as a result we got a second day's run, which is very bad, as all dimensions multiply as do your troubles.[23]

Modern Fires on the Increase

Modern fire technology has kept most fires smaller than those of presettlement times. However, fire incidence has increased with the growing recreational use since 1920. Without fire suppression, many of these fires would have burned large acreages, permanently altering the landscape in proportions much higher than would have occurred with natural fires only. Along with fire suppression, improved public awareness of potential fire hazards has slowed the increase in human-caused fires, as indicated by the increased proportion of lightning fires to those of human origin in recent years (fig. 3.9). All of these human alterations contributed to a smaller patch size in the modern forest mosaic.

The extent to which fire control has kept both natural fires and those of human origin in check led Heinselman to conclude that complete fire suppression is "almost achieved."[24] However, figures for the past thirty years reveal a startling and unexpected trend that seriously challenges this idea and changes our understanding of fire in the modern BWCA forest.

Fire frequency from all causes has actually increased about threefold during the past three decades (fig. 3.9). A similar, though less pronounced trend is also reported for the Quetico. The greatest increase has been in lightning-caused fires. This increase is far too great to be accounted for by improved methods of detecting even the small "smoke-ups" and fires that do not take off. Increased fires in recent years have also been reported elsewhere.[25]

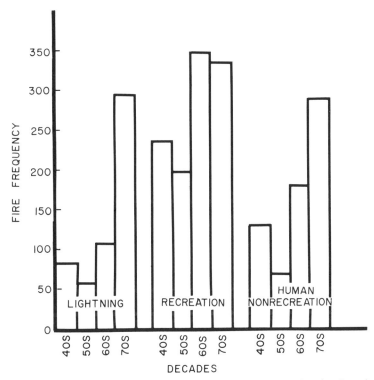

Figure 3. 9. Frequency of fires of natural and human origin by decades, Superior National Forest, 1940-1979. Data taken from Superior National Forest fire records.

Fire Size

If a fire is to have ecological significance, it must burn over enough land to be important in altering the forest mosaic. Many of the recent fires represented in figure 3.9 were small because of efficient fire suppression, calm winds, high humidity, and moist fuel. A summary of acreage burned by fires of various causes, however, actually reveals increases in recent decades (fig. 3.10). Since 1950, the acreage burned by fires of both human and natural causes has increased gradually but steadily. If the Little Sioux Fire of 1971 is charted separately, the increases are much greater for lightning-caused fires than for those originating in recreation, logging, or other human uses. In the 1970s, lightning-caused fires were responsible for more than half the total acreage burned.

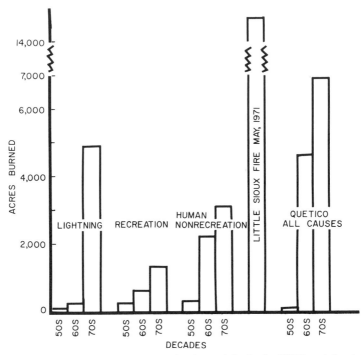

Figure 3.10. Acres burned by fires of various origins in the BWCA and Quetico
Provincial Park, 1950-1979. Data taken from Superior National Forest fire
records and from Woods and Day 1976-1977.

The BWCA acreage burned in modern decades, although not as
large as in the era before fire suppression, increased and by the 1970s
reached almost one-third that of the presuppression period, 1728-
1868. This becomes apparent when Heinselman's acreage estimates
for presettlement and for the presuppression periods are converted
into decade units and compared with recent Forest Service records.
Superior National Forest figures for average acreage burned per fire
in the 1950s was 3.4 acres; in the 1960s, 4.1 acres; and in the 1970s,
27.5 acres. Even when the large Little Sioux Fire is removed from
the 1970s acreage, the increase is still apparent.

In the Quetico, acreage burned by fires of all causes has also
increased steadily. Since more than 90 percent of the burned acreage
in the Quetico is attributed to fires of lightning origin, this represents
a trend similar to that in the BWCA. Camper fires are typical of
accessible lakeshores and islands, but lightning fires occur in remote
areas. Consequently, delay in getting crews on the ground at lightning

fires can permit significant acreages to burn before fire suppression efforts can be put into motion. Total Quetico acreage burned each decade in the 1960s and 1970s approached and may actually have exceeded that burned during any presettlement decade prior to 1860, although reports for the earliest fires may be incomplete.

Fire Stays in the Forest

Complete fire suppression, then, is far from being "almost achieved." Experienced fire fighters know that the potential for devastating fires still exists during drought years, especially since human activity has increased fire fuel, risk, and hazard. During periods of high fire hazard, many fires start and almost all spread rapidly. A ranger district may have several fires in one day, with more starting before those of the previous day are under control. Every fire must be completely extinguished or a new problem will develop. The Forest Service cannot be financed, equipped, or staffed to cope with these bad fire periods on several national forests at the same time.

> We do not need to worry that over the long haul, we will not have enough fire in the BWCA to satisfy Mother Nature's ecological needs. She will always have ample days and periods of high fire danger, severe lightning occurrence, heavy human use, resulting in many fires, plus weaknesses and breakdowns in the fire control system. Let's face it: we are not that good that we can say, when the chips are down, that we are in command of fire in the wilderness.
>
> If anyone believes that extreme drought years will not occur again in the BWCA and elsewhere in the Lake States, he hasn't seen or paid attention to the record — the repetitive relentless history of drought periods and extreme, prolonged fire danger conditions. . . . There are yet more chapters to be written on the fire history of the BWCA.[26]

Responses of Major Tree Species to Fire

Changes in the forest composition or mosaic because of fire are best understood by considering the different responses of the major tree species. Later chapters will discuss the role of logging in the direct reduction of red pine and white pine seed sources, especially in the central portion of the BWCA, and the role of white pine blister rust in destroying white pine reproduction.

Aspen often replaced the tall pines and decreased the availability of the open, postfire seedbeds needed for red pine reproduction.[27] The extension of the spring fire season by human-caused fires also favored the postfire response of aspen in the tall pine cutover sites and in places where jack pine previously dominated the postfire forest (see plate 6).[28]

Every time aspen breaks through vigorously in an opening released because of burning, logging, insect damage, or other factors, its root network obtains the vigor to push on, invading other portions ot the forest.[29] Dramatic evidence of aspen invasion into conifer forests can best be seen from an airplane during autumn foliage color change (see plate 5). The brilliant yellow projections of aspen stand out like long, invading fingers penetrating the dense green of jack pine, tall pine, spruce-fir, and cedar forests. This invasion continues and multiplies in effect. Nordin and Grigal report a threefold increase in aspen cover type between 1948 and 1970 and an 80 percent decrease in spruce-fir cover type in the western portion of the BWCA.[30]

Some botanists believe that the postfire vegetative response of certain deciduous trees, shrubs, and many common herbs is in itself an adaptation to fire, leading ultimately to the uniform distribution of these species in the forest and the smaller, patchy distribution of seed-reproducing conifers.[31] This altered distribution has been accelerated by the human reduction of conifer seed sources during logging. Most native herbs are fire-adapted in that they reappear on burned areas within ten to fifteen years after fire.[32] Actually, most fire-adapted deciduous species are really adapted to reproduce vegetatively following any type of disturbance that opens the forest floor to light and that decreases competition.

The postfire seed reproduction of white and red pine, so effective during presettlement times, was also a passive adaptation to fire—in this case, an adaptation to the fire-altered seedbed—that could have been duplicated by any other disturbance that exposed mineral soil. In the BWCA, only jack pine and black spruce, with cones opened best by heat of fire, are specifically adapted to postfire reproduction.

Any increase in deciduous cover reduces the flammability of the mature conifer forest, resulting in a gradual buildup of fuel. Aspen just does not burn as well as do the pitch-laden conifers; fires do not run in aspen stands unless there is a continuous conifer understory.

Some human influences favor the buildup of flammable fuels. Balsam fir sometimes succeeds aspen, and it is also increasing on cutover lands surrounding and within the BWCA as the two species replace the logged-off pines. This balsam fir acreage is a factor in the increasing intensity of spruce budworm attacks, as we will discuss in later chapters. Fir killed by budworm becomes fuel of explosive proportions. Many large fires of the past occurred in old stands damaged by budworm, notably the 1936 Cherokee, Frost, Rose, North, and South lakes fires and the Canadian Outbreak fires. The more recent Little Sioux Fire also raced through much budworm-killed forest. Fire destruction of balsam fir almost inevitably leads to a vigorous postfire sprouting of aspen and other deciduous tree and brush species that reproduce vegetatively.

Red pine, the most flammable border lakes conifer, was planted extensively, especially on burned land, during early reforestation efforts both within the BWCA and on adjacent lands. Many patches of planted red pine are vigorous young stands today—for example, those planted after the 1936 Cherokee, Frost, Brule, Long, and Rose lakes fires. Red pine in close, vigorous plantings is particularly susceptible to fire mortality during its first fifty years when the bark is too thin to protect the vital tissues from all but very light surface fires; the flaking quality of the bark also promotes fires running up the trunk.[33] Even without crown ignition, young red pine tops are usually killed by scorching—heat damage from hot gasses rising from burning grass, duff, and understory—at least until the trees reach a mature height. The flammability of young red pine plantations seems to invite their destruction by fire before they reach seed production age. Flash grass and bracken fires in young plantations frequently kill but do not actually burn young trees. This flammability, together with the increased danger of spread of disease and insects in extensive pure stands, has discouraged the monocultural or exclusive planting of red pine in recent years.

Jack pine is well adapted to postfire reproduction, and many stands occur in areas with high fire incidence. Fire suppression in jack pine areas leads to old age classes with increased fire hazard, thereby merely postponing inevitable fire. Survival of this species is not imminently threatened, although aspen is gradually invading the jack pine stands. If standing jack pine forests are burned and the timber is not removed, millions of seeds are released. The result is extremely heavy jack pine reproduction, "thick as hair on a dog's back." We cannot afford, aesthetically, ecologically, or economically, such stands even in a wilderness area.

Human Influences on Fire and the Changing Forest

The extent to which human activity has altered and continues to alter the forest, in terms of its flammability and postfire response, has not been fully recognized in forest management planning for the BWCA. Some plans for maintaining and restoring the presettlement ecosystem include "putting fire back" into the wilderness. This is a simplistic approach to a complex problem, a type of ecological nostalgia.

Data from recent decades show that fire has never really left the forest. Even though fires are suppressed where possible, ecologically significant ones occur and are actually increasing, as evidenced by the increased fire incidence and acreage burned by lightning fires in the past two decades.

When conditions of forest succession, disease, insect damage, stand age, fuel buildup, and weather conditions are right, the forest becomes highly flammable. All of these conditions have been influenced by human activity, as have fire frequency, acreage, season of burning, and area in which fires occur. Logging, settlement, and other human activities have altered the forest mosaic and changed the direction of its postfire response. Heinselman says that the only reason to keep fire from running in the BWCA forest is public safety and the unnatural effects caused by prior fire exclusion. However, as we have seen and will discuss in later chapters, many unnatural changes in the forest have stemmed from causes other than fire exclusion. The forests of the border lakes country are a limited acreage in which unchecked fire could do untold ecological damage.

As Woods and Day point out, for even the most fire-adapted species, there is a time *not* to burn as well as a time to burn. All plant species are fire sensitive at early stages of growth. For jack pine, this is but a few years. For the spruces, it varies with site; and for the tall pines, it is more than fifty years.[34]

The Forest Prognosis

Fire was an integral part of the original, natural forest, but the response of the modern forest to fire can never be the same as that of the primeval forest. Furthermore, the loss of thousands of acres of primeval forest to fire was tolerable because the remaining forest resource was so large. In presettlement times, the fires had little or no serious impact on human welfare or on the continent's total forest cover. Today, however, our forest resources have diminished, and we cannot possibly tolerate a fire control program that intentionally permits some large, destructive fires to burn uncontrolled in this unique wilderness area.

An understanding of past and present fire in the forest leads us to reevaluate our definition of wilderness. According to the Forest Service definition, wilderness must be "untrammeled," free from human restraints. Yet human imposition of restraints on the forest mosaic and its ability to respond to fire are very evident throughout the BWCA. Can our wilderness definition be revised to include land responding to such restraints?

Figure 4.1. Northern white cedar lob tree. Photo by authors.

CHAPTER 4

Big Cedar Lob Tree

PRESETTLEMENT FORESTS

The oldest known tree in the border lakes country is not often seen by visitors, although hundreds pass nearby every summer as they paddle from Prairie Portage to Basswood River. This northern white cedar lob tree was already at least six hundred years old when European explorers first passed the sandy bay near which it grows. The tree's exact age cannot be determined. Thin cores bored into the trunk reveal that rot has obscured the early annual growth rings, leaving the center pulpy. However, annual rings of wood laid down during the past four hundred years are sound. By counting and measuring the width of these rings and estimating the number required to fill the inner core, taking into account the possibly larger rings laid down when the tree was young, we can estimate the tree to be well over eleven hundred years old (fig. 4.1).

Considering its age and diameter, it is not tall; the bole tapers abruptly to a somewhat broken tip. Surrounding black ash and alder brush hide its thirty-five foot height from the shoreline. The thick, deeply fissured bark is further evidence of great age. One feels a sense of awe, an awareness of antiquity surrounded by transients when standing next to the massive trunk, four feet in diameter, and looking up into the gnarled branches, some more than eight inches thick. A massive root system sprawls out on the sand beneath the tree, forming a solid platform that is twenty-eight feet in diameter. This sturdy root system has been a firm foundation, protecting the tree from wind damage and keeping it upright. Mosses, herbs, and sedges carpet the platform now. Aspen sprouts have invaded it, some less than a foot from the ancient trunk. Neighboring alder and black ash fringe the margin of the root disk.

Trees of great age are scarce in this land where fire and wind have removed the forest periodically ever since the first postglacial green appeared. No other nearby cedars are more than 150 years old, so our lob tree must have lived a charmed existence and well deserves the reverence its appearance commands.

Northern white cedar is best known in the border lakes country as a swamp and shoreline species. Many lakeshores are fringed with cedars, their trunks often leaning out over the water, away from the dense forest behind them. During winter months, these fringe cedars are easily accessible from the frozen lake. Deer keep the lower branches of this choice browse pruned to an even height above the snow line, as if compelled to do so by a sense of tidiness. Away from the shore, the cedars grow straight upward, as does our lob tree.

Cedar is especially sensitive to fire damage, and it has probably survived best on shorelines and in swamps because these habitats provide some protection when fires sweep through. When conditions are right, however, it grows well away from the water on upland sites. Some ecologists theorize that it could become an upland, self-perpetuating or climax species in the unlikely absence of disturbances. Thousands of cedar seedlings one and two years old (up to twelve hundred per acre) dot the forest floor inland from the shoreline fringes. The majority of these seedlings die and are replaced by another crop the following good seed year. When grown in the greenhouse, cedar seedlings are especially sensitive to damping-off fungi, which literally strangle them at the soil line. Such fungi may also be responsible for the death of cedar seedlings in the forest. Those that survive to sapling size are often eaten by deer.

Indians and voyageurs used cedar (often called cypress or white wood) as a source of bast with which to bind and lace their canoes. It has been suggested that the name Basswood Lake for a lake surrounded by a forest containing very few basswood trees was the result of the phonetic similarity between the words "bastwood" and "basswood." The French name for this lake, however, was Lac du Bois Blanc (white wood), which could have referred to either cedar or basswood, or possibly to birch. United States Point, the northernmost point on the United States side of the lake, was once called Cypress Point, an inaccurate reference to its cedars. Frequent fires have removed all but scattered shoreline cedars from this rocky habitat. Some of these fires were set by Indians to favor blueberry growth, as described in chapter 2.

These names suggest that cedars were once more numerous in the area than they are now. The survival of the ancient tree amid much younger alder, black ash, and aspen is testimony to the transitory nature of the forests that have surrounded it in the past. In this chapter we will look at the natural forces associated with the rise and decline of the presettlement forests as a background for considering the factors influencing present-day forests.

Paleological Detective Methods

Uncovering the postglacial forest history has been a multidisciplinary task requiring tedious investigation by paleobotanists, geologists, foresters, and biologists. The story unfolds slowly as new technologies provide additional evidence.

More than ten thousand years ago, after the glaciers receded, vegetation began to develop in the border lakes country. Airborne pollen produced by that early vegetation was deposited in successive layers on some lake bottoms and in sphagnum bogs. Cold temperatures, lack of oxygen on the lake bottoms, and the acidity and antibiotic quality of peat helped to preserve this evidence.

Much of the pollen in sediment and peat layers can be identified, revealing some of the herb, shrub, and tree species present during various time periods. Occasionally, other plant parts can also be identified. We were with J. E. Potzger, one of the early pollen analysts, when he extracted cores of peat bogs in the BWCA. In one bog he recovered samples of wild rice husks from some of the deepest borings. These husks were relics from a time when the bog was a shallow, grassy rice bay, before the peat had closed in.[1]

The age of various depths and layers can be estimated by determining the carbon 14 radioactivity of the charcoal particles they contain.[2] The relative abundance of charcoal in a sediment layer is a key to fire frequency during the period when the layer was deposited. The thickest, charcoal-laden layers indicate times of increased erosion and runoff following forest fires.

Charcoal particles reflect fire frequency only in the immediate drainage area of the lake sampled, but pollen is carried for appreciable distances by air currents before it sinks to the ground. Consequently, pollen in the sediment of an isolated bog or lake can be an indication of vegetation that occurred elsewhere in the area. For example, the abundant ragweed pollen found in layers laid down in the late 1800s is believed to reflect settlement and agricultural activity to the south and west. Airborne pollen settles on lakes and becomes incorporated in lake sediment; pollen from insect-pollinated species does not. Consequently, pollen profiles give clues only to the wind-pollinated species, not the total vegetation. However, the pollen of most trees, common brush species, sedges, grasses, and some herbs are air borne. Growth requirements of species identified in the samples are clues to moisture and temperature conditions occurring at the time the layer was deposited.

Climate and the Ancient Forests

Charcoal fragments found in layers deposited more than 10,000 years ago indicate that fires were common at that time, when the climate was cooler and more humid than it is now. Pollen samples reveal a tundra and open woodland with sedges, grasses, herbs, and scattered spruce covering much of the area. About 10,000 to 9,000 years ago, red pine, jack pine, and some birch pollen were also deposited. Gradually, the climate warmed and became drier. The dryness and increased flammable vegetation contributed to more frequent fires. The warming trend peaked 6,000 to 7,000 years ago, and at that time white pine was more abundant that jack or red pine. Northern white cedar was also common. Small amounts of pollen from tree species that now grow farther south were also found in samples taken from layers deposited during this period, so we must visualize some pockets of hemlock, walnut, hickory, oak, and elm scattered throughout the area.

The climate has been cooling gradually since that time. Samples deposited about 3,000 years ago reveal a gradual decline in abundance of white pine pollen and an increase in jack pine, spruce, and alder pollen. Layers deposited 3,000 to 1,200 years ago contain less charcoal and pine pollen, but alder and cedar pollen are more abundant. This period is believed to have been cool and damp with fewer forest fires than before. Between 1,200 and 500 years ago, our lob tree grew from a seedling to be a tree more than 500 years old. During that time, fire frequency increased slightly; the vegetation remained fairly stable, but it included more jack pine and less cedar. It is estimated that many forests burned at intervals of 10 to 100 years, giving us even more reason to marvel at the longevity of our cedar lob tree.[3]

From 1550 to 1840, the so-called Little Ice Age was associated with decreased fire frequency, a slight increase in spruce, and a slight decrease in pine. Farther north in the Quetico, similar trends have been detected, although the pine dominance was not as great and declined more rapidly. Spruce has always been more abundant farther north.

A picture emerges of vegetation developing gradually under control of the climatic change that followed the retreat of the glaciers. During that drying and warming period, early tundra species slowly gave way to open spruce stands and eventually to pine. With the increase in more flammable, resinous conifers, forest fire became a more frequent influence on developing vegetation. But the climate, with its alternating periods of drought and abundant rainfall, also influenced fire frequency. Other factors—blowdown, disease, insects, species interaction, and competition—all played roles in shaping the forest at times when it was particularly sensitive to them. These factors also often influenced the response of the forest to fire and climate.

The cool, moist period brought a decline in acreage of larger pines. About 300 to 250 years ago (the late seventeenth century) when Europeans arrived, the forest had developed into a mosaic of spruce-fir, jack pine, birch, aspen, red pine, and white pine, with other species in lesser quantities.

The Presettlement Forest Seen by Explorers

As mentioned earlier, some of our knowledge of more recent pre-settlement forests comes from the fire history revealed in scars on older trees, the ages of stands, and charcoal in upper soil layers (fig. 4.2). Notes, paintings, and drawings made by early explorers and the records of the first surveyors also contribute to a more vivid picture. Many early explorers and surveyors kept journals filled with fascinating tales of hardship, danger, and the sense of excitement that comes only when facing a vast unknown. Since their primary interest was in water routes through the area, land form, and orientation, comments on the forests were fragmentary; vegetation was apparently mentioned only when it hindered their progress or when nothing else about the landscape was noteworthy.

Early French explorers visited the border lakes in the 1600s when our lob tree cedar was more than 800 years old. After that, expeditions into the area gradually became more frequent and extended. Forest fires and charred landscapes were often described in early diaries. In 1735, Jean Pierre Aulneau, a Jesuit priest, tells of traveling all the way from Lake Superior to Fort St. Charles (Lake of the Woods) unable to see the sun because of smoke. He believed those fires had been set by Indians, but he offered no proof.[4]

Figure 4. 2 Charcoal in an upper soil layer giving evidence of past forest fire.
Photo by authors. Originally published in T. T. Kozlowski and C. E. Ahlgren,
Fire and ecosystems (New York: Academic Press, 1974).

Travelers in the first half of the nineteenth century included
cartographers, geologists, and a few naturalists. David Dale Owen
commented in 1848 on the fire-scarred land found in places and
noted that the aspen, fir, spruce, pine, and birch between Lake
Superior and Lake Vermilion were small, recent reproduction.[5] Paul
Kane's journal from the same period is the only one to mention
forests severely defoliated by "green worms," presumably the forest
tent caterpillars that periodically plague aspen stands today:

> It was a remarkable fact that the trees . . . for a full 150 miles of our
> route were literally stripped of foliage by myriads of green caterpillars,

which had left nothing but bare branches; and I was informed that the scourge extended to more than twice the distance I have named, the whole country wearing the dreary aspect of winter at the commencement of summer. As it was impossible to take our breakfast on land, unless we made up our minds to eat them dropping incessantly as they did from the trees among our food, and the ground everywhere covered with them en masse, we were compelled to take it in our canoes.[6]

Dr. John Bigsby, a physician and artist traveling with the International Boundary Commission expedition in 1823, wrote one of the more graphic descriptions of the area.[7] He mentioned pine, cedar, spruce, and aspen forests, occasional burned land, and dense brush underwood. He also observed streaks of yellow scum (conifer pollen) dried on rocks and cliffs along the lakes, indicating changes in water levels, and he noted that the larger the lake, the greater the water level fluctuation. Tree species can often be identified by their characteristic shapes in his drawings. One picture shows a distant shoreline on Lac La Croix, obviously denuded by fire.

Poor Dr. Bigsby should be commended for his graphic descriptions and pictures because his notes are well sprinkled with astonished and agonized comments on the ever-present, ferocious mosquitoes. He must have traveled during a particularly bad "bug" year, for he indicated that the travel party was forced to wear gloves, veils, and caps over their ears and to tie their pantaloons down over their boots. He described a voracious mosquito that continued sucking blood even when its abdomen or blood sac and wings were removed.

The First Border Lakes Country Survey

Notes made by surveyors working for the General Land Office give us our clearest picture of the border lakes presettlement forest. The original survey was begun in 1858 and delayed by the Civil War, but by the late 1800s it included 750,000 acres of the present BWCA. Unlike early explorers, the surveyors moved inland from the lakes and streams, ran lines through bogs and black spruce swamps, scrambled up rocky hills and through dense brush. J. Wesley White, Superior National Forest historian par excellence, has sorted out the terse statements of these men concerning the rugged, inhospitable nature of the forests they traveled. A few of these statements bear repetition as a record of one more wilderness concept:

In summing up the entire contract, I may truly say that it lies in the most forsaken country it has ever been my misfortune to encounter.

There is apparently nothing (here) that would induce a sane person to enter within the unsacred domain of moose, wolves, bear, snow, rain, mosquitoes, flies, rocks, swamp, brush, and rapids. . . .

In describing the lands embraced within this township, the best expression will be found in a long sigh ending in a groan. . . .

The mosquitoes and "no-see-ums" have the snow and water at an advantage for punishing a human being. Life is almost unendurable from the torture of insects and physical discomfort. I have been stung by mosquitoes in this township while standing in snow knee deep. . . .

The memory of a man in the relentless grasp of a pack sack loaded with 100 pounds of provisions and camp plunder, on a June day, midst myriads of mosquitoes, deer and moose flies, and no-see-ums is sufficient to drive a man to the depths of perdition. . . .

We have frequently . . . exhausted the last pound of eatables and travelled a day or more without breaking our fasts. On one occasion, a single pigeon supplied a corps of three men during three days.[8]

In spite of hardships and starvation, these men made detailed notes on each township and section. Although a few European settlers had already entered the area, their effect on the forest was minimal; most forests surveyed originated long before the arrival of Europeans. Much of the land had probably not even been penetrated by Indians, voyageurs, or explorers, for their numbers were small and they all stayed close to the water routes. Commercial logging was still in the future. Surveyors' notes, therefore, give a good indication of the natural forests growing around our cedar lob tree just before the major impacts of modern civilization.

The surveyors were acutely aware of the dense undergrowth, rough topography, and desolation on the more than half of the land that bore signs of recent fire. Most of them had moved north after establishing section lines on treeless prairies, and they were far from enthusiastic about the border lakes wilderness. As one surveyor noted, "Altogether, this township comprises as undesirable piece of land as I have ever seen."[9]

The surveyors knew nothing about the prehistoric fire origin of the forests they encountered. They judged entirely by readily visible evidence. It is not surprising that on fifteen townships, or almost half of the boundary waters land surveyed, no signs of fire were detected and the land was recorded as unburned. The charcoal record and most recent fire studies have shown that most of the surveyors' "unburned" townships had actually burned at some time in the past but had been free of fire for more than a hundred years, long enough to produce mature forests. Eight of these townships contained high-quality red and white pine. Six others had heavy timber or pine without species designation.

Aspen, fir, birch, and occasional cedar and larch were mentioned as understory or as occurring in scattered patches on almost 90 percent of the forested townships. Jack pine was recorded as an associated species in only six townships, a good indication that fire had been absent from much of the area for some time. This absence of the fire species strengthens the paleological evidence of a Little Ice Age, when fires would have been less frequent. Aspen was recorded on eleven of the fifteen townships, and it was probably a well-distributed but subordinate part of the tree cover.

The abundant blueberries, raspberries, and currants noted in one township suggest a recent fire undetected by surveyors. Another township was found to have little timber except brush, young aspen, and jack pine. In the absence of evidence of fire, this type of cover could also have followed a recent blowdown. Surveyors noted some signs of past fire on eighteen of the thirty-three townships. They described fire-scarred trees, fire-blackened land, young forests with charcoal on the ground, and other fire signs still obvious. The fires had occurred mostly in "large timber," "old pine," "white pine," or "red and white pine." Only one spruce-fir stand with scattered birch and white pine was recorded as recently burned.

Surveyors were primarily interested in establishing corners, however. They could not have known the intensity or season of the fires. We cannot be sure of the species involved in places where they recorded merely "timber" or "large timber." However, the old red and white pine were not reproducing themselves immediately after fire, for no very young stands were noted even though ecologists speculate that there had been at least twenty-five generations of white pine in the area during the previous 6,000 to 8,000 years.[10]

Aspen-birch and spruce-fir were the most frequently mentioned forest types, followed by jack pine and brush. Quantitatively, aspen-birch reproduction was reported in 90 percent of the townships, spruce in 40 percent, jack pine in 40 percent, fir in 33 percent, and dense brush in 28 percent. Although it would be wrong to read too much about forest succession between the lines of their bug-stained logbooks, their observations were sufficient for F. J. Marschner to use to construct a map of the original vegetation of Minnesota.[11] This map will be referred to in detail in chapters 6 and 7.

Some evidence of the size and extent of the pine forests can be found in photographs taken in or near the border lakes country in the early 1900s. One, taken on the road to Burntside Lake from Ely

Figure 4.3. Road from Ely to Burntside Lake, 1909, showing dense forest of
towering pines dwarfing the horse-drawn buggy. Courtesy Lee Brownell.

in 1909, clearly shows the dense forest of massive pines dwarfing the
horse and buggy on the new dirt road (fig. 4.3).

The available evidence, then, shows that in the late nineteenth
century the BWCA forest was a mosaic of old red and white pine,
spruce-fir, aspen, birch, jack pine, brush, and a scattering of other
species. Distribution of these species had been guided over the
centuries by the receding glacier, changes in rainfall and temperature,
and the associated varying patterns of recurring fire. Each factor—
glacier, climate, and fire—appears to have dominated in the control
of vegetation at different times, but all were important in shaping
the pattern of the total presettlement forest.

Figure 5.1. Paper birch lob tree. Courtesy Larry Ahlgren.

CHAPTER 5

Paper Birch Lob Tree

EARLY INHABITANTS

Our paper birch lob tree stands among other birches on a gentle slope just above the shoreline (fig. 5.1). Throughout the open, sunny grove, most birches stand in clusters of two to six trees that are joined at the ground line or standing close together. The trees in each clump originated as sprouts from the base of an older tree, now long gone. Punky remains of parent stumps can still be detected at the bases of younger birch clusters. The many birch clumps covering the slope indicate that an earlier birch generation grew here about one hundred years ago, the old, deteriorated trees gradually giving rise to this new stand of clustered trees. Birch is a short-lived species that is old at eighty years, but it regenerates quickly, often in association with aspen, spruce, and balsam fir.

Elsewhere in the area, groves of single birch not in clusters are also common. The single-standing growth habit indicates that they originated from the light, wind-disseminated seeds often seen scudding over snow in winter.

Farther back from the shore, the white bark on a number of trees is interrupted by a rough, corky, brown band about six inches wide where the outer bark has been removed. In earlier times, Indians removed birch bark in the spring for use in constructing canoes, wigwams, baskets, packs, and even shrouds because it was waterproof, pliable, and lightweight. Early inhabitants did not have to go far for this remarkable material; birch could be found throughout the forest. In the spring when the cambium (the inner growth tissue that separates bark from wood) is moist and tender, sheets of bark can be peeled off easily. On trees with straight, branchless trunks, pieces four or five feet long and two feet wide can be obtained. The tree will die if it is completely girdled when the bark is removed, especially if the bark is removed down through the cambium or if large pieces of bark are removed and the inner, delicate tissues are exposed to the sun and dried out. If the cambium survives on at least part of the tree's circumference, the tree will continue to grow; however, the damaged bark will be brown and corky, never regaining the original smooth, white quality.

Paper birch, more than any other tree, stands as a symbol of the ways in which border lakes country inhabitants of the seventeenth, eighteenth, and most of the nineteenth centuries adapted their life-styles and livelihood to the available forest and water resources. Birch is not a large tree; it did not dominate the landscape, but its bark provided transportation and shelter for both Indian and European inhabitants. Our consideration of human impact on the wilderness ecosystem must begin with those early forest dwellers, for their lives reveal interesting patterns of wilderness use with which we can compare later, heavier human impacts.

How It All Began

This was a harsh country, sustaining only a small Indian population. From prehistoric times until the late seventeenth century, the area was used lightly by the Dakota-Assiniboin Sioux. Shortly before Europeans arrived, however, Ojibway moved in from the east under pressure from the Iroquois nation and the expanding European settlements. This shift in Indian populations reflected different tribal preferences and capabilities to adapt to a changing forest ecosystem. Even more important, the Ojibway migration illustrated how insignificant events in the expanding European culture produced domino effects of great significance to the future of North America.

In the border lakes country, the domino sequence began well before 1600 on the Atlantic coast. Every year, a few adventurous Breton, Norman, and Basque fishermen crossed the Atlantic to fish the Grand Banks. Whenever they landed to repair nets and boats, they traded with the natives, taking beaver, fox, and ermine pelts back to Europe.

These furs, obtained under cold winter conditions, were of such excellent quality that merchants willingly paid large sums for them. They far surpassed the furs obtained in more southern parts of North America where newly forming colonies also began carrying on active trade in furs. Northern beaver pelts were especially prized. Their fur was long and silky with a very thick wool beneath, ideal for making fine quality felt to fill demands caused by lack of central heating and the high fashions of court life. Other furs were also popular.

Soon Champlain, Henry Hudson, Radisson, Groseilliers, and other seventeenth century explorers began to investigate and returned to Europe with impressive accounts of New World fur resources. French and English trading posts and settlements sprang up in Quebec, on Hudson Bay, and eventually farther south and west.[1] Competition between the French and English for fur trade with the Indians led to open warfare.[2] Many books have been written about this period of North American history, but we will concern ourselves only with the events directly related to the border lakes history and ecology.

A Westward Migration

The powerful Iroquois nation befriended the British when they arrived. The French, therefore, considered the Iroquois to be enemies as competition and hostilities between British and French increased. Under pressure from the French, some of the Iroquois moved into territories of less powerful Algonquin tribes, pushing them west.

Among the Algonquin tribes displaced by the Iroquois were the Cree and Ojibway, who moved into the northern lakes regions of what later became Minnesota, Wisconsin, Michigan, and adjacent Canada. The Cree soon moved on north and northwestward. Many of the Ojibway remained, however, gradually pushing the resident Sioux south, west, and northwest. By 1750, the Ojibway dominated the border lakes, except for sporadic Sioux war or hunting parties and other small conflicts that occurred during the next one hundred years.

Tribal Differences and Forest Ecology

It might seem puzzling that the traditionally hostile Sioux gave up the northern lake country to the more peaceable Ojibway with only an occasional skirmish, when they ferociously resisted intrusion elsewhere. While some Ojibway had guns and the Sioux did not, this lack of firearms did not intimidate the Sioux in other places. Bloody Sioux raids on white settlements and battles to retain central, southern, and western portions of Minnesota are well known. Attacks and massacres occurred wherever the Sioux came in contact with European settlers, traders, or explorers. However, by the eighteenth

century the forests of the border lakes had changed since the prehistoric times when ancestors of the Sioux first laid claim to them.
As we will see, these forest changes, together with the different life-
styles of the two tribes, are possible reasons why the Sioux departed
with only minor contest.

When the Sioux dominated the area there had been many dry
years, a long period of drought with extensive forest fires throughout
the Lake States. Tree ring experts date this drought at over five
hundred years ago. Evidence from border lakes pollen and charcoal
sediments, discussed in chapter 4, push this warm, dry period back to
about twelve hundred years ago. During this dry period of high fire
frequency, extensive patches of early postfire sprouting vegetation
would have been developing—a woodland type providing a good
food supply for large, browsing animals such as moose, caribou, and
deer. Any human culture dependent on these large animals for food
and other life necessities would undoubtedly have had good hunting
for many years. The Sioux had just such a culture.

During this time, extensive postfire pine forests also originated
in what later became Wisconsin, Michigan, and Minnesota. Such
forests take more than a century to mature and can remain healthy
for several hundred years if left undisturbed. After between seven
and eight hundred years, the dry period was followed by the cool,
moist Little Ice Age that occurred between 1500 and 1850. Fires
became less frequent, and many more pine stands escaped fire than
previously. These stands grew to maturity and became the big pine
later seen by surveyors and logged in the early 1900s. They will be
discussed in chapter 6.

Because old pine forests provide less browse for large mammals,
hunters would find game less abundant as the forests matured. Life-
styles would have to adapt accordingly or other hunting grounds be
found. Differences between Sioux and Ojibway in their dependence
on large animals for food, shelter, and transportation influenced
their enthusiasm for the changing border lakes forest conditions of
the seventeenth century.

When the Sioux found water travel necessary, they usually used
dugouts or hide-covered small craft. Dugouts were time consuming
to produce, heavy, and hard to portage; hide-covered craft were
awkward, easily torn on rocks and snags, and required skins of large
animals for their construction. The Sioux were not canoe builders;
the canoes they had were obtained from Ojibway by fair means or
foul. The Ojibway, however, entered the area by canoe and were
master canoe builders, using birch bark laced with cedar bast or

Figure 5. 2 Ojibway squaw and children in birch bark canoe near Indian village on Lac La Croix, 1915. Photo by William Magie. Courtesy U.S. Forest Service.

spruce roots and sealed with spruce pitch to produce and repair these graceful craft. All necessary materials could be found on almost any lake shore. If animal hides were available, the birchbark was sometimes covered with hide for cold weather travel. Canoes were well adapted to travel in the area's many lakes and rivers, and they were light enough for easy portaging (fig. 5.2).

The Ojibway lived most of the year in small family units and moved from place to place with the season and the food supply, hunting, gathering food, finding wintering grounds. Most families wintered by themselves, and the families came together as a tribal or village unit only for a time in spring and early summer—from sugar bush (late April) to berry ripening moon (July). The canoe gave Ojibway great mobility for a roving life-style in lake country. With small units and canoe travel, the tribe spread its use of the forest over a broader area than did the Sioux, who lived mostly in small villages from which they sent out hunting parties.

The Ojibway originally lived in dome-shaped wigwams made of woven rush mats and birch bark, banked with boughs and snow in winter (fig. 5.3). Birchbark wigwams made Ojibway existence in the maturing forest less of a struggle than that of the Sioux, who preferred hide-covered teepees and were thus dependent on large animal skins for shelter (fig. 5.4). Even with a maturing pine

Figure 5.3. Typical Ojibway wigwam covered with birch bark near Burntside Lake, 1916. Photo by William Trygg. Courtesy U.S. Forest Service.

Figure 5.4. Ojibway adaptation of Sioux style of teepee, using birchbark covering instead of skins, near Basswood Lake in 1913. Photo by William Trygg. Courtesy U.S. Forest Service.

forest so prevalent, birch could always be found in the border lakes country; large animal hides, however, had become difficult to obtain.

Although both tribes hunted and fished, their food preferences also differed. At the time of European contact, the Ojibway diet consisted of fish, game, berries, wild rice, and wild herbs. The journals of European travelers mentioned the Ojibway using fish as food much more often than game, a possible indication that fish were more readily available at the time. Alexander MacKenzie, traveling for the fur trade in the late 1790s noted that game was scarce and that Indians subsisted on fish and wild rice. They told MacKenzie, however, that in the past both game and Indians had been more numerous. In 1793, John MacDonell of the Northwest Trading Company wrote in his diary that his party found only fish for food on the entire route from Grand Portage to Rainy Lake. Several other nineteenth century explorers lamented the lack of wildlife found during their travels. Dog meat was a delicacy and was cause for a feast.[3]

The Ojibway were skilled at smoking and drying fish. Packed in birchbark containers sewn shut with spruce roots and sealed with pitch, the dried fish kept well for winter use, supplementing pemmican (dried meat) when game was scarce. The Ojibway actually preferred fishing for a livelihood—to such an extent that fur traders soon learned to save fishing nets for only the choicest barter. If fishing were made too easy, the Indians would stop hunting for fur-bearing animals.

Ojibway pictographs of moose reflect the tribe's fascination with the unusual sighting and killing of one of these large animals, ranking them close to the legendary monsters like Misshipeshu. Fish were everyday, staple items that were rarely drawn. The fact that some pictographs show both moose and deer suggests that they were probably drawn in the late 1800s, since deer were infrequent before the area was logged.

The Sioux fished sporadically, but their diet was more dependent on large mammals than was that of the Ojibway. Pemmican was a staple for them. As the forest matured and browse diminished, moose and caribou became scarce; a diet depending on large mammals was in jeopardy.

The tribes also differed in their response to European contacts. Coming from the east, Ojibway brought with them guns, metal tools, cooking pots, and woven blankets, all obtained from Europeans in exchange for furs. These things increased their creature comforts and, in certain situations, assured their actual survival. The Sioux, on the other hand, were still essentially living in a Stone Age culture through the eighteenth century; they had few if any metal tools, guns, or blankets. The Ojibway's early acceptance of trade items (and unfortunately also their passion for alcohol) made them eager to barter with fur traders who began plying the border lakes shortly after the Ojibway arrived. The Sioux, however, were slower to accept most European wares; most often they fought or retreated in the face of European contact.

Although better adapted to the northern forest lake land than the Sioux, the Ojibway population fluctuated and was never large. Disease, harsh winters, and occasional periods of famine kept populations low. In 1807, for example, Dr. John McLoughlin, manager of the Northwest Trading Company, recorded fewer than five hundred Indians living in the entire district from Rainy Lake to Grand Portage.[4] If the population of any large mammal were so low today, it would be classified as an endangered species. It has been estimated that before European settlement, the entire Indian population of Minnesota never exceeded fifteen thousand. Currently, over one million visitor days are recorded in the BWCA alone each summer, in sharp contrast to that meager native population of the seventeenth through the nineteenth centuries.

Distant European pressures and the changing forest thus both played roles in shaping the Indian population. The Sioux were not willing or able to adapt to the maturing forest conditions and decreasing wildlife populations of the eighteenth and nineteenth centuries, and they were also hostile to European influences. Their use of the area was light anyway, and they were readily replaced by the Ojibway, whose food, shelter, and transportation preferences better fitted the forest conditions of the time and who accepted European contact.

Arrival of the Voyageurs

How those early Basque fishermen would have marveled at the chain of events begun by their casual barter during their trips to the Grand Banks! Not only were two Indian tribes displaced, but demands for fur continued to increase rapidly. Cutthroat competition led to French and British trading posts on Hudson Bay, the St. Lawrence, at Quebec, and farther west. The Hudson's Bay Company motto, Pro Pelle Cutem (skin for skin), is an indication of the intense competition.

Profits of over 600 percent were common. One beaver pelt was worth twelve buttons; a blanket cost six pelts, a gallon of brandy cost four pelts. By 1713, Britain's powerful Hudson's Bay Company controlled all the posts in the Hudson Bay area. Independent Montreal fur traders, hard put to obtain fur, began sending voyageurs inland to trade with the Indians. The Indians, glad to avoid the long annual trip to a trading post, learned to wait for the traders to come to them. Thus the voyageur traffic began. Other trading companies were formed, and soon all were pushing farther west in vicious, sometimes bloody competition. The fur trading companies began establishing trading outposts farther and farther inland, bringing more and more traffic to the border lakes routes.

In 1768, Grand Portage on the short of Lake Superior became an assembling post for voyageurs. Each spring, they gathered there on the edge of the border lakes country and organized into groups of about six men. Packing their trade goods and meager supplies into a twenty-five-foot birchbark *canot du nord,* each group would head into the *pas d'en haut,* across the border lakes to Lac La Croix, Rainy Lake, and points beyond. Some hardy men overwintered along the route, establishing trading outposts varying from small log cabins to La Vérendrye's elaborately palisaded Fort St. Charles on Lake of the Woods. In 1861, trading posts were still functioning on Basswood, Saganaga, Mountain, Vermilion, La Croix, and Rainy lakes. Moldering log bases of cabins can still be found on these and other lakes along the route. It is impossible to determine whether these old cabins were used by voyageurs or Indians, however, because many Ojibway adopted the white man's log cabin for winter use. We have found trading beads, buttons, and bottle fragments in soil on some of these cabin sites.

As competition for furs increased, more and more voyageurs passed through the border lakes in midsummer on their way into the forest, returning laden with furs when ice left the lakes the next spring. Many groups only passed through the area to and from large outposts on Rainy Lake, Lake of the Woods, and as far north and west as lakes Winnipeg and Athabaska. They were often accompanied on such trips by *bourgeois* (company officials) and *commis* (clerks). Traveling parties of twenty or more were not uncommon.

Records of the many competing traders and companies are only partially preserved, so it is impossible to estimate the total volume of fur removed from the area during the peak of the fur trade era. In 1746, a trader named Dobbs brought out 49,000 beaver skins. Even beaver castoreum was prized for medicinal purposes. Between 1858 and 1884, Hudson's Bay Company obtained 25,000 pounds of the material, selling it for ten to twenty-five dollars a pound. Since five or more male beavers are needed to produce a pound, this quantity alone would represent over 125,000 male beavers, and not all traders bothered with this product. Indeed, the Indians often kept it for themselves.[5]

Whatever figures appeared on the fur trade company books, there is little doubt that heavy trapping reduced local populations of fur-bearing animals. However, the beaver decrease was apparently much greater than could have been caused by trapping and natural population fluctuations. Grace Lee Nute cited devastating fires in 1803 and 1804 as additional reasons for beaver decline,[6] although Heinselman reported no evidence of major fires those years.[7] Beaver are able to obtain food from larger birch and aspen trees than can large browsing animals. The beaver decline came later than the decrease in hide-bearing large mammals associated with reduced browse in the maturing forest.

For whatever combination of reasons, fur was no longer plentiful in the area by 1804. Grand Portage was closed and the major voyageur route shifted northward, running from Fort William through the Kamenistikwa route, although some trading posts remained open in the lower border lakes for another half century (fig. 5.5).

In 1826, the Ojibway ceded the mineral rights in Minnesota and adjacent territory by accepting terms of the Treaty of Fond du Lac. Commissioner McKenney introduced the treaty to assembled chiefs in Duluth by saying, "I note that in all your great country there is little beaver; that your woods and streams are silent, that but little game of any kind can be found, and that your traps are slow to snap."[8] The fur-bearing animals were not completely eliminated, however, and with managed, restricted trapping most species have gradually increased. Only the wolverine is believed to be extinct in the area; cougar and marten are rare.

Only fragmentary evidence of other direct ecological impacts of voyageurs can be found. Some writers believe that the even-aged pine trees, 100 to 140 years old, that were cut by the first loggers about 1900 originated following fires caused by voyageurs. Studies indicate that some major fires occurred in the latter part of this period, but they could have been of natural origin. At present, it is estimated that beaver dams and subsequent flooding cause 14 percent of the total mortality of balsam fir and spruce in the area. We can therefore assume that beaver or the lack of them made at least a temporary mark on the forest ecosystem in the past.

The overwintering voyageurs were the first European inhabitants

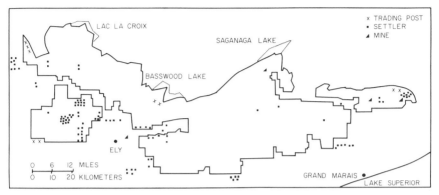

Figure 5.5. Centers of European activity before 1870 in the region now comprising the BWCA and adjacent counties. The map shows approximate locations of fur-trading posts, settlers' cabins, and mines.

of the border lakes, other than occasional exploration parties. Except for the time they spent at larger trading posts, the voyageurs adopted a forest life-style much like that of the Indians. Their direct impact on the forest was correspondingly light. In their quest for furs, however, they required the Ojibway to harvest a forest product in greater quantity than the Indians would have obtained for personal needs. Indians had no informed scruples against heavy trapping. Thus, although voyageurs and the trapping may have had only small and temporary ecological impact, they marked the beginning of utilization and exploitation of the area's resources—one more domino toppled in the sequence.

The importance of the border lakes as a travel route for fur trade, exploration, and migration is reflected in the Webster-Ashburton Treaty of 1842, which established the border between Canada and the United States. The route running along the border and cutting through the tip of Minnesota's Arrowhead to Grand Portage on Lake Superior was so important that the treaty decreed that it must forever remain open to citizens of both countries. This consideration is still honored today, despite increased regulation in surrounding lakes. The final section of the route, Grand Portage, is outside the BWCA. It is the only strip of land in the world entirely within one country but freely open to citizens of another.

European Settlers

A few European settlers arrived in the late 1700s and early 1800s, most establishing themselves near fur trading posts. Settlement increased after 1854, when the Treaty of LaPointe gave the Ojibway rights to sell land and threw the Minnesota Arrowhead region open to development. Whether the settlers came primarily for trapping, prospecting, or cutting timber, they all needed at least some cleared land. To hold land, homesteaders were required to erect a building and give some evidence of living there. It was not until the Timber and Stone Act of 1878 that timber and mining interests could buy undeveloped land and hold it for later use.

The first recorded agricultural development was the settlement established in the mid-1700s by Lord Selkirk on Fowl Lake. Records are scanty, but we know that the settlement failed and the buildings were deserted by 1823 when John Bigsby saw them.[9] Lord Selkirk was part owner of the Hudson's Bay Company and later established trading posts and settlements at Rainy Lake, on the Red River, and much farther west. Presumably, even in his early days with the company, he felt the need for settlement. An agricultural establishment near the busy fur trade community at Grand Portage would have been useful because food staples had to be hauled by *canot de maître* through the Great Lakes all the way from Montreal.

Surveyor notes from the late 1800s tally at least 81 settlers' establishments within the area now included in the BWCA. By the time of the survey, however, many cabins had already been abandoned, but twenty-seven sections were still occupied by settlers; seventeen of these included enough cleared land to merit a note by the surveyor. In township 64, ranges 14W and 15W, for example, west of Burntside Lake, sixty-eight settlers with cleared land, houses, and outbuildings were noted.[10] Another 150 cabins were located in townships outside of but adjoining the present BWCA borders. Many of these were also abandoned by the time of the survey (fig. 5.5).

Most settlers took land in the eastern or western parts of the present BWCA or in the separate portion south of what is now the Echo Trail. The central section, containing much of the large pine that was heavily logged less than half a century later, was left unclaimed.

Access was probably a determining factor in the settlers' locations. Entry to the eastern section via Grand Portage was possible. The northwestern section could be entered via the Dawson Portage route that extended from Lake Superior to the Red River via Sturgeon Lake, Maligne River, and Lac La Croix. This route was used by hundreds of immigrants moving farther west in the mid-1800s. Portages and trails were traveled with oxen, horses, and wagons; launches, tugs, and barges plied the lakes. Dawson route traffic, even that of people who only passed through, served further to open the area. Possibly a few of the early border lakes settlers originally planned to go farther west, but for reasons long lost they decided to try their fate in the northern lake country.

Most early settlers avoided the pine forest and built their cabins on burned-over land or land noted by surveyors as "aspen-birch." Only two abandoned cabins were found on land the surveyors recorded as "large pine." Young, postfire aspen and birch required less ax work than big pine in those days before chain saws when a dense, tall pine forest would have seemed dark and threatening to settlers. Imagine the high hopes of early pioneers when they saw the lush growth in the first, small gardens they planted on recently burned land! The postfire fertilizing stimulus of ash lasts only a few years, however, and later crops became successively poorer.[11] This decreasing fertility and other hardships made it impossible to alter the forest ecosystem to fit an agricultural life-style.

Unlike the Ojibway and fur traders, these settlers were looking for a resource that wasn't there—fertile soil. Indeed, the U.S. Soil Conservation Service classifies soils of northeastern Minnesota as distinctly nonagricultural. Any resident attempting to garden without using fertilizers, organic material, and extra water during dry summers can attest to this. Small wonder, then, that settlers intent on hay fields and planted crops abandoned their holdings after a few years. Homesteaders fared somewhat better on similar land elsewhere in northeastern Minnesota only because access roads permitted them to augment their incomes by working in logging camps and mines. Even so, the many deserted, crumbling farms at the end of brushed-in trails are testimony to the lack of fertile soil.

The impact of early settlers on the BWCA land was probably slight, but they, their oxen, horses, and baggage undoubtedly introduced some of the nonnative plants present today. The small cabin sites were the first of many private holdings in the area. Such private ownership eventually made the BWCA unique among wilderness areas and complicated its administration.

Search for the Mother Lode

Prospectors also arrived, seeking still other resources—gold, silver, and copper. They established exploratory pits and a few small mines, which they soon abandoned (fig. 5.5). Most gold and silver prospecting occurred in the eastern portion near the Gunflint Trail. One reported silver vein on Crystal Lake was uncovered but never worked. Farther west, the famous 1866 gold rush up the Vermilion Trail to Lake Vermilion did not penetrate the present BWCA. The gold rush did, however, bring many people into adjacent areas and increased population and human movement around the wilderness. Prospecting led to the ultimate discovery of massive iron deposits adjacent to the border lakes country. Subsequent iron mining held people nearby and brought many more. Copper resources were not found in the early days and have not yet been mined.

Gold and silver prospecters and miners were not successful. Few, if any, worked their claims for more than a year or two. They, like the settlers, sought a resource that was not there. Their direct impact on the forest ecosystem was small, but like the settlers they laid claim to land, owned portions of it, and tried to alter the forest rather than adapt to it. They were harbingers of what was yet to come, one more domino toppling in the wilderness.

The BWCA retained its forests and remoteness through those early days only because its resources were not suitable for agricultural or mining expansion. The border lakes country was a land where, as the fur traders found, heavy harvest or use of a resource soon depleted it. Human use during the prelogging days left little permanent evidence of trammeling, but it set the stage for future use. It was a land to which settlers and miners brought the still contested idea that wilderness could be bought and possessed as private property.

Figure 6.1. Eastern white pine lob tree. Photo by Lawrence Berg. Courtesy *American Forests.*

CHAPTER 6

White Pine Lob Tree

PINE LOGGING

When they first look up into our towering white pine lob tree, most border lakes visitors describe it as "Majestic!" No other word so aptly suits this 325-year-old monarch with its straight, proud trunk towering 125 feet skyward, well above most of the forest around it (fig. 6.1). Its graceful branches catch the breeze, translating it into a soft whisper—the murmuring pines, according to Longfellow. Among the branches, one or two stand out, dead and bare or with browning needles. These dead and dying "flags" indicate that our lob tree is infected with white pine blister rust.

The rust arrived in the area in the early 1900s when the lob tree was already mature; it was not killed by the disease, losing only an occasional branch to it. Young trees are more susceptible to rust than are their aged parents, however, and on the forest floor beneath this monarch the fate of its offspring is another story. Tiny seedlings and slender saplings two to four feet tall hint of a future white pine forest. However, few intermediate-sized pines can be found. Before most young trees grow above a man's head they develop rust cankers on their mainstems, are strangled, and die.

Other old red and white pine grow nearby, part of an imposing stand. Similar Pine Points, Pine Islands, or Cathedral Groves are sparsely scattered throughout the area. Such stands, along with the occasional old white pine trees amid younger forests of other species, are living remnants of the pine forest early surveyors recorded in the area. Crumbling pine stumps two to four feet in diameter can still be found throughout woodlands in northern Minnesota, as ghosts of those tall pine forests of presettlement days.

Our towering white pine lob tree, then, is a fitting reminder of the original logging industry and its effects on border lakes forest succession. The harvest of white pine throughout New England and the north central United States was critical in the development of our nation, laid the foundation for the economic and industrial growth of Minnesota, and changed the course of forest development in over half of the BWCA.

White Pine and the New Nation

To understand the full impact of the removal of this species, we must go back to colonial days when the white pine harvest began gathering momentum in New England. Straight eastern white pine trunks excelled any timber found in Europe and were prized for masts of schooners, clipper ships, and ships of the realm. The British Crown soon established exclusive rights to prime white pine. By 1761, all mast-sized white pine were reserved for the British navy, and severe restrictions were placed on colonial logging operations. These restrictions triggered a spirit of independence among the colonists as they protested vigorously their loss of freedom in the forests.

The cutting restrictions did not mention Indians, however, and colonists defiantly donned Indian garb to continue their cutting of the resource they so badly needed for trade and expanding settlement. Thus, the mood and garb of Boston Tea Party participants in 1773 originated with the loggers a decade earlier. White pine became the symbol of independence on the first flag of the pre-Revolutionary Forces in 1775, and seven states still retain it in their flags or state seals.

As the new nation expanded westward, the vast white pine forests were assumed to be inexhaustible; timber was cheap and plentiful. White pine lumber industry became the first industrial giant in much of the country. The smooth, strong, easily worked wood was in heavy demand for building homes and furniture, looms and factories, bridges and freight cars, schools and churches. The massive lumber industry moved rapidly and relentlessly from New England through Pennsylvania and New York, then on to Michigan, Wisconsin, and Minnesota. Large red pine often grew in association with white pine and was harvested with it. The tall pine forests of these two species, like the virgin topsoil and minerals, were exploited in the relentless drive to build the nation and expand its economy. These were the days when wilderness was no longer considered hostile; it was an untapped source of riches, free for the taking. Railroads that followed the loggers brought more settlers to the recently cleared land, and the nation continued to grow and push westward.

Minnesota's White Pine Industry

From 1839, when the first Minnesota sawmill opened on the St. Croix River, until 1932, when most of the last large mills had closed, 67.5 billion board feet of pine lumber were removed from Minnesota forests. Minnesota pine formed the economic foundation for the state's iron and grain industries and provided the lumber for building major cities in other states—St. Louis, Omaha, Kansas City, Des Moines, and other settlements along the Mississippi River and westward into the prairies and plains.

Toward the end of the nineteenth century, pine in readily accessible Minnesota forests along the Mississippi and its tributaries had been cut and sawmills along the rivers closed. Railroads were built to remove sawlogs from more remote areas.

By 1890, most of the state's remaining tall pine was limited to the Arrowhead region of northeastern Minnesota, including the border lakes country.[1] The tall pine extended in an irregular band north, east, and west of Duluth. One large finger of tall pine reached into the central portion of the present BWCA, with the finger's tip ending in the Quetico, less than ten miles north of Basswood Lake. Other, smaller sections of tall pine were found in the eastern and western parts of the present BWCA. Logging operations north of Duluth began in various places. Here, as elsewhere, railroads accompanied the logging (and mining) operations. Roads in the Ely area were short and limited. The first trip by automobiles from Ely to Duluth, for example, was not made until 1910; it was accomplished largely on old tote roads with the extensive use of ax and shovel. Railroads provided the most important link with the outside world. Even without actual cutting within the BWCA area, the logging, along with iron mining, would have brought many people to live next to the BWCA.

Logging in the Border Lakes Forests

Paul Bunyan yarns seem but slight exaggerations when compared with the records of the two major logging companies operating in the central portion of the border lakes country from 1893 until the late 1920s. During twenty-five years of logging in the area, Swallow and Hopkins Company built over fifty camps, each averaging about one

Figure 6.2. Sunday at a Swallow and Hopkins logging camp on Washington
Island, Basswood Lake, in the early 1900s. Courtesy Lee Brownell.

hundred men and sixteen horses (fig. 6.2). Beginning in 1898, the
company operated a railroad across the Four Mile Portage from Fall
Lake to Ella Hall and Mud Lakes, and by 1901 it had extended it to
Basswood Lake (fig. 6.3). Between 1898 and 1911, the company
hauled over 300 million board feet of pine by rail over the portage.
Steam tugs and "alligators" (figs. 6.4 and 6.5) supplied the camps
and hauled log rafts down the lakes to the Winton sawmill (fig. 6.6),
one of the largest in Minnesota. Other rail spur lines were later built
from Fall Lake to Pipestone and Jackfish bays of Basswood Lake
and to Gun Lake.

The Company logged in the Burntside, Fall, Newton, Basswood,
Moose, Wind, Ensign, Crooked, and Knife lakes areas—portions later
destined to be the most heavily used for recreation. Swallow and
Hopkins harvested about 600 million board feet from their border
lakes forest holdings, although the company estimated that somewhat less than half of their 200,000-acre forest holdings contained
enough pine to be operational. [2]

So extensive were the rafts of logs that older Ojibway in the
area recall the heavy deposits of pine bark chips that settled in
shallow bays of Basswood Lake. This bark inhibited the growth of
wild rice, seriously reducing the harvest for over a decade. [3] The
effect was temporary, however, and wild rice again thrives in shallow
bays along the loggers' old routes.

Figure 6.3. The Basswood-Fall Lake railroad over the Four-Mile Portage, 1901. Photo by Will Jeffery. Courtesy U.S. Forest Service.

Figure 6.4. The tugboat *Merti J* operated by Will Jeffery on Basswood Lake during the early 1900s to supply logging camps and to pull log rafts. Courtesy U.S. Forest Service.

Figure 6.5. An "alligator" approaching Hoist Bay, Basswood Lake, with raft of logs, early 1900s. Note man with pike pole on raft. Courtesy Lee Brownell.

Figure 6.6. Swallow and Hopkins sawmill, Winton, Minnesota. Photo by C. E. Scarlett. Courtesy U.S. Forest Service.

   The St. Croix Lumber Company was the largest in the border lakes country. The company had moved north after depleting its pine holdings in east central Minnesota. Between 1896 and 1923, its total harvest from the area was a billion board feet. The company cut in the central portion of the border lakes including Basswood, Burnt-side, Fall, Kekakabic, Frazer, and Thomas lakes and the Fernberg, North Kawishiwi, and Stoney River areas that are well known to present-day canoeists. Other major loggers included the Pigeon River Lumber Company and the General Logging Company, which operated in the eastern section, and the Vermilion-Rainy River Company, which operated in the western section between 1910 and 1925. Some smaller companies were also cutting.

   Much of the area to be logged was covered with pine and could be clear-cut—that is, all or most of the trees removed. Where good pine was mixed with other species or with younger pine, the forest was high-graded—selectively cut of good pine only. The few young stands were left uncut. Some ecologists believe that the bulk of the pine harvest was done by high-grading,[4] but old-timers recall that high-grading was usually limited to holdings adjoining clear-cut operations, so-called rubber forties or illegal extensions of cuttings on the edge of good pine stands. It would not have been economically feasible to establish a cutting operation in an area for high-grading only.[5] A close look at the pine stumps remaining in the logged area, pictures of the pine harvest (fig. 6.7), and logging company cutting records confirm the heavy cutting and large size of many of the trees cut.

   The forest was not entirely tall pine, however. Most of the patches of spruce, balsam fir, and deciduous trees were left uncut. Some birch was cut for spur line railroad ties. Other wood was used to build logging camps, corduroy access roads, and bridges. The wood-burning rail and boat steam engines used for hauling and towing also required a steady supply of fuel wood.

   Bustling mill towns appeared on the edge of what was to become the BWCA. Some men worked in the woods during the winter months when logging camps were most active and when much of the cutting and hauling from the woods was done. When the spring log drive was completed, they returned to the mills to work through the summer months.

   Iron mining, begun in the mid-1800s, had also drawn many people to the area. The population surrounding the border lakes grew rapidly and brought other human ventures into the wilderness. Commercial fishing enterprizes appeared on Crane Lake, Lac La

Figure 6.7. Typical load of white pine removed from the border lakes forests near Winton, early 1900s. Courtesy James C. Ryan.

Figure 6.8. Leo Chosa operated a commercial fishing business on Basswood Lake until 1916. Photo by W. H. Magie. Courtesy U.S. Forest Service.

Croix, Rainy Lake, and Basswood Lake. Some fish companies supplied the Lake Vermilion gold rush settlement that temporarily mushroomed in 1894. Others provided fish for logging camps and stores. The Booth Fish Company of Duluth regularly obtained lake trout, whitefish, walleyes, and bass from Basswood Lake (fig. 6.8). For a few years around 1909, a thriving market for loon skins developed at Grand Marais. Loons were shot on border lakes and their skins used in coats, sleeping bags, and feather skin robes to be shipped to the Hudson's Bay Company and Alaska.[6]

Commercial hunters supplied camps and local butcher shops with moose, venison, and occasional caribou. Game was apparently plentiful as recovering vegetation on cutover land provided increased browse. In 1909, L. R. Beatty, one of the early rangers, reported counting fifty-seven moose one afternoon while traveling the five to six miles between Lakes Agnes and Nina Moose.

In February 1909, a meeting of the North American Game and Fish Protective Society was held in Toronto. Its purpose was approval of General Andrews' plan to establish a game preserve in the border lakes country in an effort to save moose from extinction. That same year the Minnesota Game and Fish Commission established the Superior Game Refuge of over a million acres, and the Superior National Forest was established.

By 1902, Pacific coast timber began competing with Minnesota pine on the lumber market. In 1910, the first rangers on the recently formed Superior National Forest reported very little tall pine left in the border lakes country. Mostly jack pine remained, and it was considered a useless weed tree at the time. Sawmills began closing in the late 1920s.

When the last large mills closed in the early 1930s, the forests of the canoe country had undergone dramatic changes. A comparison of acreages of various kinds of forest in the original geologic survey and the 1948 Forest Service timber survey indicates that before logging, tall pine covered 334,080 acres (522 sections). After logging, tall pine remained on only 26,560 acres (41.5 sections) or 8 percent of the original tall pine forest. These figures compare well with those compiled by the North Central Forest Experiment Station, with additions from the Superior National Forest Timber Atlas, which reveal that 321,480 acres were logged before 1940. As a result, one-third of the area, primarily in the central portion but with additional tracts in the west and east, was cut over and interlaced with logging roads, trails, camps, and occasional railroad spur lines (fig. 6.9).

Although some remnant tall pine stands escaped harvest, most

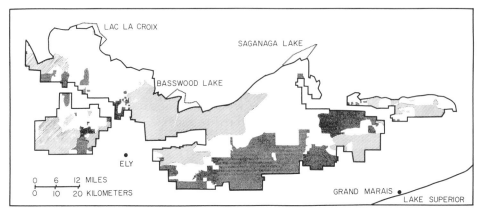

Figure 6.9. Timber harvest in the BWCA. Dark-shaded area: land logged after 1940, primarily for pulpwood; light-shaded area: land logged prior to 1940, mostly for red and white pine; unshaded area: land not cut.

of the uncut, so-called virgin pine forest remaining consists primarily of the smaller, short-lived jack pine that do not produce prime lumber and had been rejected by loggers.

A distinction between jack pine forests and the taller red and white pine forests is frequently omitted in figures and discussions of modern pine forests remaining in the BWCA. Consequently, the public erroneously visualizes tall pine whenever the terms "virgin pine" or "pine forest" are used. The two forest types are much different aesthetically as well as ecologically in response to both natural and human disturbance.

Ecological Change after Logging

Removal of the tall pine was a turning point in the ecological history of the area. Prior to the logging era, red pine and white pine had reseeded and grown up anew following repeated, devastating wildfires for six thousand to eight thousand years, or more than twenty-five generations. Ancient fires swept through the forest with such regularity that few stands survived longer than our lob tree, which is some three hundred years old. Heinselman's maps of early forest fire history reveal that before logging, most tall pine forests had not burned since before 1610.[7] In contrast to such natural disturbance, the pine harvest was thorough, abrupt, and concentrated in the tall pine forest, laying the foundation for new and far-reaching ecological change. It also accelerated natural changes that were proceeding slowly prior to logging.

Although loggers removed 92 percent of the mature tall pines, some old white and red pines are still present. However, there is a striking lack of young trees, aged thirty to eighty years, established after logging. Throughout the BWCA, the 1948 timber survey found natural red pine reproduction on only five sections, or less than 0.5 percent of the area. White pine reproduction was found in small patches adjacent to old seed-bearing trees on only sixteen sections, or less than 1 percent of the area. As we shall see, most of this reproduction has a slim chance of survival. Thirty- to eighty-year-old white pine trees do not occur in the Quetico.[8] Some ecologists still predict that tall pine forests will be reestablished naturally, as they were before logging, in spite of the lack of young, postlogging stands. Present evidence makes this possibility appear remote.

Silvics and Postlogging Forest Development

The tall pine trees do not regenerate readily in today's forest because their species characteristics and growth requirements do not adapt readily to present-day conditions of seedbed, competition, and disease. Far too often we speak of the forest only as a collective unit, forgetting that it consists of several living species, each with well-defined requirements that limit its ability to respond to the environment.

Red pine occurs occasionally in pure stands, especially on islands and along lakeshores where competing brush is scarce (fig. 6.10). Its acreage is more limited than that of white pine, but the two species are often associated in mixed stands on drier sites. Red pine is genetically the least variable native pine species, as can be seen in even-aged plantations where the trees are uniform in height, growth rate, and form. This lack of variability makes the species ecologically inflexible and therefore less adaptable than the other two native pines.[9]

During the Ice Age red pine was virtually eliminated, except for a few refugial populations located primarily in the southeastern part of the United States. A few small, outlier populations outside the present natural range of the species still exist in West Virginia and Illinois. Present-day northern red pines are believed to be derived from one such population. If this is true, red pine from that one inbred stand would have completed migration north after the glaciers retreated, only about eight thousand to eleven thousand years ago. Pollen in sediment profiles document the appearance of red pine in the border lakes country at about this time. Not enough time has passed since then for the development of genetic variation or

Figure 6.10. Naturally occurring, young red pine on brush-free shoreline with thin soil, 1919. Photo by A. H. Carhart. Courtesy U.S. Forest Service.

heterozygosity through the slow process of accumulation and re-combination of natural mutations.[10]

Lack of genetic diversity makes red pine unable to adapt to changing conditions. The species originally occupied about 7,500,000 acres within its natural range. It now grows naturally on only 1,500,000 acres. Some pine specialists believe that, like other plant and animal species with genetic inflexibilities, red pine is on the way out in the natural forest, although it is by no means a threatened or endangered species.[11] In addition to its natural stands, extensive planting has increased its total acreage.

Both red and white pine have east-west ranges from the East Coast to eastern Manitoba, but red pine's north-south range is less than half that of white pine, extending only to central Minnesota, Wisconsin, and New Jersey (except for those few outlier populations elsewhere). Red pine has a much smaller altitudinal range than does white pine, although this is generally not a limiting factor within the flat Lake States.

Young red pine stands did not develop after logging because red pine reproduction requires a strict, rigid set of conditions that were not met on cutover land. The first condition is an abundant seed supply. Red pines begin producing cones when about twenty to twenty-five years old, with heavy crops only every six to ten years. Even in a good seed year, this species produces much less seed per tree than does white pine, and after logging the supply was drastically reduced.

Secondly, red pine seeds germinate well on exposed mineral soil or thin moss, but they are sharply inhibited by ash on new burns. Accordingly, seedlings do not appear on burned areas until ash on the soil surface has been washed and leached by rain and melting snow for several years. Germination is also inhibited by birch, balsam fir, and big-leafed aster litter. Seeds may germinate on thick moss or deep pine litter, but young roots are unable to grow through these mats and become established in the mineral soil beneath, a process necessary for obtaining a constant supply of water and minerals.[12]

Finally, red pine seedlings require open sunlight for good growth and survival. Hazel, alder, raspberry, bindweed, or heavy grass usually appear when old vegetation is removed by cutting or natural disturbance and the forest floor is opened to light. Such vegetation is made especially lush and rank by the minerals released by the ash and by the increased sunlight reaching the ground when overstory has been removed.[13] Red pine trees do not grow well in the shade of this vegetation, and they are usually choked out. Even in more open areas, herb competition for moisture may inhibit early stages of red pine growth until brush or aspen close in. For these reasons, attempts to establish red pine by direct seeding of cut or burned areas are rarely successful.

Reduction of seed supply by logging drastically reduced the probability that a good seed crop would occur and be dispersed to an area free from brush and low vegetation exactly when the seedbed is in good condition. Consequently, red pine regeneration is largely restricted to open, thin litter sites, usually along lakeshores and on islands. Occasional postlogging red pine clusters seventy to eighty years old can be found on such border lakes sites today. One stand exists on Washington Island, Basswood Lake, on a site logged in the early 1900s. Most young stands were planted, however, as we shall see in the next chapter. A similar condition exists in Itasca State Park farther west in Minnesota, where long-term studies have concluded that red pine will not reproduce naturally and will require the help of management techniques.[14]

In contrast to red pine, white pine seems to be well adapted to reestablishment after logging or any other disturbance. Its wide natural range, extending from Newfoundland and eastern Manitoba southward to southern Iowa, Kentucky, Georgia, and the Appalachians, attests to its ecological flexibility. This species varies in size, shape, growth rate, and disease resistance, indicating that it possesses the genetic diversity necessary to adapt it to a variety of conditions.

White pine produces cone crops much earlier than does red pine; often at ten years of age. Heavy crops occur at intervals of three to ten years; branches bend low like those of heavily laden apple trees, almost unable to support the weight of the prodigious cone crop in bumper seed years.

White pine seed germinates well on fresh ash, mineral soil, and most types of forest litter. Even on thick needle and moss seedbeds, the young roots grow vigorously and reach mineral soil successfully. Seedlings grow well in open sunlight, but they are also shade tolerant and can survive under competing vegetation.

White pine seedlings and saplings are common in cutover portions of the border lakes forests. However, trees established after logging that are twenty-five to eighty years old are even less frequent than are red pine. Their absence is the result of both tall pine harvest and the well-meaning attempts to undo the damage of that exploitation. Once again, we can follow the domino effect of early human pursuits elsewhere as they shaped the destiny of both people and tall pine forests in the border lakes country.

White Pine Blister Rust, an Alien Threat

To discover why white pine did not return after logging, we must go back to prior events in the eastern forests of North America. Shortly before 1900, after the white pine harvest had swept across most of the northeastern United States, people began taking a long, hard look at the vast acreage of devastated, cutover land. Often ravaged by fire, the cutover was unattractive and unproductive. Loss of the white pine industry had already hit the white pine states where it hurt, in the pocketbook, and growing public opinion was that "something should be done about it." Soon everyone—the Forest Service, state conservation agencies, schools, ladies' groups, churches, youth organizations—were planting trees. If white pine was the tree that had been harvested, white pine was the one to be put back wherever possible. School forests, church forests, county forests, state and national forests were all heavily planted with white pine.

Figure 6.11. Fatal aecial canker of white pine blister rust on young white pine. Photo by authors.

American tree nurseries, accustomed to coping with modest white pine orders for parks and civic plantings, were unable to meet the sudden increased demand for white pine planting stock. European nurseries, especially in Germany, could supply them cheaply and in quantity, and soon millions of European white pine were imported and planted in the eastern United States and Canada. Some of them were infected with white pine blister rust (*Cronartium ribicola*), a disease previously unreported in North America but long known in Asia and Europe (fig. 6.11). It causes only minor damage to closely related European pine species, but it is often fatal for eastern white pine.

In 1897, blister rust was first found in the United States at Kittery Point, Maine, on English stock of European black currant, one of the alternate hosts of the disease.[15] Other introductions of the disease on white pine stock from Germany and France followed. The disease was found on white pine in Philadelphia in 1905 and in New York the following year.[16] Undoubtedly several shipments contained infected seedlings, for blister rust spread rapidly through cutover lands wherever young natural or planted white pine seedlings were found.

White pine blister rust requires the presence of an alternate host plant belonging to the genus *Ribes* (gooseberries and currants) to complete its life cycle and spread the disease. As luck would have it, several *Ribes* species thrive on cutover and burned-over land and were abundant wherever white pine grew. Within a few years, spread of blister rust reached alarming proportions. Dead and browning seedlings and saplings added to the devastation on cutover lands and new plantations.

Under pressure from foresters, Congress passed the Federal Plant Quarantine Act in 1912. The first plant quarantine banned importation of white pine seedlings. Although quarantine was later effective for control of other plant diseases, the act locked the barn door after blister rust was already racing over white pine lands.

By 1915, blister rust was found in Minnesota, and by 1933 it was present and increasing wherever white pine grew in the state. It was first reported in St. Louis County in 1919 and in the Tower area adjacent to the present BWCA in 1920. However, many of the photographs taken by Arthur Carhart during his recreational evaluation of the area in 1919 show mature white pine with the flags indicating damage by blister rust. The disease was thus prevalent in the central portion of the border lakes country less than twenty-five years after it was first discovered on the North American continent (see fig. 6.21, page 118).

Records of early rust damage are incomplete. However, on 250 acres of white pine planted between 1922 and 1935 halfway up the Gunflint Trail (T65N R2W), 1940 survival was less than 1 percent, giving strong evidence of rust efficiency. In 1934, only 1 percent of the natural white pine was infected with rust in the eastern portion of the canoe country, according to a Forest Service survey; by 1940, infection had reached 31 percent with over 60 percent of the alternate host, *Ribes,* infected. Tree rings in wood beneath rust cankers indicated that initial infection in this portion of the BWCA must have occurred about 1920. Since Carhart's pictures showed rust in the Kawishiwi (west central) portion in 1919, it is possible that rust moved from west to east along the border lakes, carried by westerly winds and air currents during the spore-producing season. Peak rust infection and spread occurred around 1935, and by 1940 the disease was declining slowly as the number of host trees was reduced.[17]

Today, white pine grows rust-free in warm, sandy areas where

Figure 6.12. Branches killed by blister rust (flags) on old white pine. Older trees are less susceptible to rust than are saplings, usually losing just the tops or some branches. Photo by authors.

Ribes is not abundant. Postlogging, "old-field" white pine that has reached maturity is now being cut in woodlots and forests in New England. Rust hazard is now low in Michigan, Illinois, and southern Wisconsin, but is higher in the north.[18] The cool, moist climate of northern Minnesota is well adapted to the spread of rust spores, and the disease still eliminates most white pine reproduction today. In Minnesota, rust hazard has been rated on a scale of 1 through 4, with St. Louis, Lake and Cook counties of the Arrowhead region rated as 4, the highest hazard. Rust hazard in the counties surrounding Minneapolis and St. Paul is very low, with a rating of 1.[19]

Mature trees that escaped logging were already beyond the highly susceptible seedling-sapling stage before blister rust became prevalent. They are seldom killed, but many have dead tops or branches (fig. 6.12), as does our lob tree. Tops killed by blister rust in older stands permit the entrance of rot and boring insects and hasten the decline of the few relic old stands that would otherwise survive until blowdown, fire, or lightning strikes destroy them.

In 1915, an extensive campaign was begun to eliminate all *Ribes* plants from white pine areas. This grew to be the largest forest disease control program ever attempted. In Minnesota during the 1930s and 1940s, hundreds of workers from the Civilian Conservation Corps, Works Progress Administration, Forest Service, and state agencies tramped the forest, acre by acre, pulling and destroying *Ribes* bushes. Some *Ribes* eradication was done in uncut and cutover sections of both eastern and western border lakes country and in the Basswood Lake area, between 1934 and 1950.

One of the most extensive projects was in the eastern area, following the international boundary from north Fowl to Watap and Rose Lakes, bounded on the south by John, McFarland, Big Caribou, and Little Caribou lakes (fig. 6.13). Here, amid both cutover and uncut white pine lands, a crew of fifty men established a camp and began work in 1934, the year rust was discovered in the area (fig. 6.14). Eradication was planned for large portions of the forest, and undoubtedly many of these plans were at least partially completed. However, *Ribes* quickly resprouted from any crown fragment left in the ground, and because the work was not halting the spread of rust it was abandoned in the early 1950s.

White pine is genetically more diverse than red pine; among the resultant variations, a few trees produced in the natural forest are able to resist rust or survive it with minimal damage. Rust resistance is not widespread in the natural forest, however, so the spontaneous development of an extensive white pine forest that is resistant to rust cannot be expected. However, by locating resistant trees, breeding them, testing the offspring for resistance, and selecting promising survivors, resistant white pine seedlings can eventually be produced in nurseries and returned to some northeastern Minnesota forests for outplanting. The Wilderness Research Center, in cooperation with the Forest Service and the University of Minnesota, is doing such work and is establishing experimental plantings of second-generation hybrids selected for rust resistance on the Superior National Forest, state, and industry-owned land (figs. 6.15-6.20).[20]

Decline of the Tall Pines

Destruction of vast seed supplies by logging, the ecological inflexibility of red pine, and the introduction of white pine blister rust have all contributed to reducing the tall pine forest in the canoe country to a fragment of its former acreage. The optimum conditions of seed source, open seedbed, and lack of competition under which these species reproduced naturally for more than twenty-five generations no longer exist. As a result, forest development, dominant vegetation,

Figure 6.13. Portion of one of the largest blister rust control areas in the border lakes, located in the eastern part of the area. This view shows East Pike and Big John lakes. Courtesy U.S. Forest Service.

Figure 6.14. Big John blister rust control camp, 1939. Photo by E. E. Honey. Courtesy U.S. Forest Service.

Figure 6.15. A naturally occurring, disease-free white pine is selected for use as breeding stock for resistance to blister rust. Photo by authors.

Figure 6.16. Branch from the selected tree is grafted on planted rootstock in a seed orchard. Photo by authors.

Figure 6.17. White pine seed orchard in Duluth, Minnesota. Photo by authors.

Figure 6.18. Pollination bags protecting cones are used in controlled pollinations. Photo by authors.

Figure 6.19. Vials of pollen from selected disease-free trees are preserved for future pollinations. Photo by authors.

Figure 6.20. Hybrid white pine seedlings (F_2) for outplanting and testing for rust resistance. Photo by authors.

forest age distribution, fuel load, and the patterns and timing of forest fires in the area have all been altered.

Concern for extinct, endangered, or threatened animal and plant species is prevalent today. Rare and declining species are carefully monitored and protected. We know from the past that without protection they will vanish. The passenger pigeon, for example, was once common in the United States and was reported in the border lakes country as late as 1885. They disappeared, presumably because promiscuous harvest reduced the population to such a perilously low level that one unreported natural disaster or disease could and did eliminate the remaining population. Certainly, the tall pines do not fit into this category, for both species grow naturally and are planted elsewhere. However, a realistic understanding of the modern wilderness forest must recognize the decline of these species. It must also acknowledge that this decline has been hastened—and in the case of white pine, directly caused—by human pursuits within the area and elsewhere on the continent.

The fate of certain pine species elsewhere provides precedents for accepting the human role in the decline of the tall pines. In Ireland, south England, and Denmark, natural stands of Scotch pine no longer occur, although peat bogs in these countries contain many stumps indicating past Scotch pine forests. Experts believe the pine was systematically removed, beginning in the Stone Age. Portions of the forests were cleared, cultivated or grazed for a time, and then abandoned as people moved on and cleared other areas. Scotch pine did not return to the cleared land for reasons about which ecologists can only speculate now. Other species, including oak, birch, heath, and bog plants replaced the pine wherever cultivation did not maintain fields and downs. Scotch pine, of course, thrives naturally elsewhere and is also planted extensively, but in these particular regions its ability to reproduce naturally was destroyed by primitive clearing.[21] Many centuries of timber harvest and overgrazing have virtually eliminated other forest tree species throughout the Mediterranean area.[22]

An Altered Forest Mosaic

At the present time, changes in forest composition caused by the sharp decline in tall pine reproduction are not yet obvious to the casual canoeist. The new forest on former tall pine lands consists of native species already common in the area. However, the proportion of each species in the patchwork or mosaic has been changed in subtle but important ways. These changes are evident only when we compare the distribution of modern forest types with that of the pre-logging forest. Fortunately, evidence for such comparisons is available.

In 1930, F. J. Marschner prepared a map showing the presettlement forest condition. He used as evidence the General Land Office Survey notes obtained by those first hardy surveyors described in chapter 4.[23] The survey was completed just before the tall pine harvest reached into the border lakes country. Recent ecological research and Superior National Forest timber surveys reveal the present condition. The most accurate timber survey was completed in 1948; recent surveys are based in part on that of 1948 and are incomplete. Marschner's map does not have all lakes accurately plotted and has a higher total land acreage than do recent surveys. Only federally owned lands were included in the recent timber surveys. Because of these differences, comparisons made are in percentages rather than total acres.

Marschner recorded four major upland forest types in the border lakes country: (1) white and red pine forests (only one patch of about a thousand acres between Ensign and Kekakabic lakes was classified as pure white pine, the rest as predominantly white pine with red pine associated); (2) jack pine; (3) aspen-birch (conifer), which included all lands dominated by aspen and birch but with significant elements of balsam fir, spruce, and young pine that could eventually replace the hardwoods; and (4) aspen-birch (hardwood), with little evidence of conifer reproduction and probably destined to remain as hardwood, mostly aspen, birch, and red maple.

The 1948 Forest Service survey also distinguished between tall pine and jack pine types. However, increased knowledge of forest succession, more detailed information, and presence of later successional stages because of fire suppression made possible a subdivision of aspen-birch (conifer) into classes with various proportions of pine, spruce, and fir. For purposes of comparison, we have combined all aspen, birch, spruce, and balsam fir categories in both surveys.

A summary of the two surveys reveals changes in proportion in the upland forest mosaic:

	Prelogging Survey (Marschner Map)	1948 Timber Survey
White pine-red pine	36%	4%
Jack pine	21%	36%
Aspen-birch-spruce-fir	43%	60%

The decline in the percentage of tall pine was accompanied by a corresponding increase in jack pine and the aspen-fir sequence, suggesting that these latter types invaded the former tall pine forest. A 1944 reconnaissance of the Roadless Area by Forest Service personnel confirms this interpretation:

> It was apparent that there had been a gradual transition from red and white pine type to jack pine, upland spruce-fir, and aspen-birch types. The remains of *heavy* red and white pine were very evident under well-stocked jack pine in the 65 and 90 year old age classes.[24]

A count by quarter sections of presence of absence of these species in the 1948 timber survey further describes this invasion. Regardless of forest type designation, aspen was present on 72 percent of the quarter sections, spruce-fir on 60 percent, jack pine on 52 percent, white pine on 15 percent, and red pine on 8 percent, demonstrating the invasive nature of aspen, with balsam fir following. White pine reproduction of sapling size was found on less than 1 percent and red pine reproduction on less than 0.5 percent of the quarter sections, including pre-1948 plantations.

Changes in percentages of the various forest types within the mosaic become even more evident when cutover acreages are compared with uncut lands. Forest types of a nine-township (more than 200,000 acre) portion of uncut forest in the central BWCA is compared with an adjacent cutover portion of the same size:

	Uncut Portion		Cutover Portion	
	Marschner Map	1948 Survey	Marschner Map	1948 Survey
White pine-red pine	38%	1%	73%	3%
Jack pine	35%	56%	10%	34%
Aspen-birch-spruce-fir	26%	43%	17%	63%

In the presettlement forest as revealed in the Marschner map, the white pine-red pine type was not as extensive in the portion later left uncut as in the area that was to be harvested. Tall pine harvest was, of course, concentrated in the central portion that contained more tall pine (fig. 6.9). Almost half of the uncut forest was predominantly jack pine and was therefore passed over by early loggers as not profitable to harvest. In contrast, the cutover area originally contained only 10 percent jack pine. By 1948, the tall pines had declined in both the cut and uncut areas, and aspen-birch-spruce-fir increased correspondingly.

Figure 6.21. Veteran white pine tree in 1919 that survived fire thirty years before. The surrounding postfire vegetation is dominated by jack pine. Note the blister rust flags on white pine at this early date. Photo by A. H. Carhart. Courtesy U.S. Forest Service.

The Uncut Forest

Jack pine had increased noticeably in the uncut forest by the time of the 1948 timber survey (fig. 6.21). The original jack pine areas had been maintained and increased to include some former tall pine lands. Fire, blowdown, insects, and disease were the major disturbances. Since jack pine is well adapted to reproduce following fire, it can be assumed that fire played a significant role in that increase. Aspen-birch-spruce-fir also increased, including some on former tall pine lands with more moist, deeper soils than those invaded by jack pine. By 1948, most of the original aspen-birch-spruce-fir areas contained mixtures in which spruce-fir dominated, although some aspen patches were also found.

The Cutover Forest

Although tall pine forest originally covered more than 70 percent of the cutover land, cutting decreased it drastically to 3 percent. Unlike the uncut area, the greatest increase on cutover land was in the aspen-birch-spruce-fir sequence, with aspen-birch dominating. This deciduous species domination contrasts sharply with the composition of the aspen-fir sequence in the uncut area where spruce-fir dominated in 1948. As in the uncut forest, some of the drier tall pine sites were invaded by jack pine following fires. Jack pine invasion of tall pine land was not as great as on uncut lands, however, probably because of heavier, moist soil and brush competition.

The Contemporary Condition

During our thirty-five years in the central cutover pine area around Basswood Lake, we watched the gradual decline of the first post-logging aspen forest in the early 1950s, its replacement by understory balsam fir, the budworm-initiated decline of balsam fir in the 1970s, and a resurgence of aspen that is just now becoming evident. For more than twenty years, we have recorded the vegetation on over two thousand permanent plots distributed throughout the BWCA. On these we have seen aspen regenerate dramatically following spruce budworm infestation, wildfire, prescribed burning, logging, and blowdown on both coniferous and deciduous sites. In contrast, balsam fir reproduction is sparse, many seedlings perish, and only 2 percent live for fifteen years or longer.[25]

With each replication of the cycle, aspen acreage increases and conifer acreage decreases. For example, in 1954 Heinselman determined that only 7.5 percent of Minnesota's aspen stands would eventually convert to balsam fir.[26] This low rate would be expected because aspen-birch survives well on a wider range of soil moisture conditions than does balsam fir.[27] The resilience of aspen is slowly and methodically having its effect. What might have happened slowly and naturally over thousands of years in some portions of the forest has been accelerated and expanded as a result of logging.

The twentieth century changes that we have studied reflect but a brief moment in the total chronicle of forest succession. However, these changes are of a quality and proportion that would not have occurred without human intervention. Knowledge of the growth requirements of the forest tree species dominant today help us understand the present status of those species and also give clues to their future in the constantly changing forest mosaic.

Another Lesson in Silvics

The fire adaptations of jack pine are more specialized than those of the tall pines, giving it an advantage whenever fire occurs. At the same time, these fire adaptations virtually limit jack pine reproduction to burned-over lands. Jack pine begins producing cones when only three to seven years old. Most cones are hard and woody, remaining closed on the tree and containing viable seed for up to twenty years. The cones are quite fire resistant, but heat of fire causes them to open, dispersing seeds on recently burned land.

Jack pine seedlings produce short, slow-growing roots that become established best on mineral soil, especially where fire has removed the humus layer. Germination is not inhibited by ash, but the litter of deciduous trees, balsam fir, and red pine inhibits germination and early growth.[28] The young seedlings are even less shade tolerant than red pine seedlings and must become established during the first few postfire years before other vegetation shades the seedbed. Once well established on open land, the seedlings grow rapidly, competing successfully with other developing vegetation.

Aspen and birch, with their prodigious supplies of windblown seed, can become established on sites opened by fire and other disturbances. The seeds and young seedlings are easily dehydrated, however, and mortality is very high except on moist sites. Most successful aspen and birch reproduction is vegetative. Birch sprouts from the bases or stumps of older trees, especially when the older trees die or are cut or when the forest floor is opened to light. Aspen suckers from shallow, lateral roots that sometimes stretch eighty to one hundred feet away from the parent tree.

The ability to produce suckers begins when aspen seedlings are still small and continues throughout the life of the tree. One root system can support fifteen or more suckers at one time. These, in turn, can resucker, resulting in a widespread, interconnecting clone. A few trees or seedlings in or near dense stands of other species can maintain a root system capable of resuckering abundantly when disturbance occurs. Roots can remain alive for three or four years even if the trees or suckers from which they originated are removed. Removal of older trees stimulates suckering, as does opening the area to light by any of a variety of disturbances, including fire. Since suckers grow rapidly, often three to five feet during the first growing season, they compete successfully with associated low vegetation and dense brush.

Aspen, which occurred as occasional stands and scattered trees in the presettlement forest, can now be found in most upland forests of the BWCA. Although often associated with jack pine on recently burned lands, it is able to thrive on more varied sites and is therefore more universally present.

Balsam fir, black spruce, and white spruce, the remaining major elements of the forest mosaic, are more shade tolerant than the other species. Consequently, they can become established as understory in pine and aspen stands and will be the dominant species when the overstory breaks up. Balsam fir produces heavier and more frequent seed crops at a younger age than do the spruces, and it is more prevalent. White spruce, however, can grow to a greater age, become taller, and dispatch seed for greater distances than can balsam fir.

Aspen and balsam fir often replace each other in a repeating sequence. Balsam fir and spruce become established as a shade-tolerant understory in aspen stands; when the aspen declines, the conifers dominate. When in turn the balsam ages and declines, more light reaches the ground and the almost omnipresent aspen root network is stimulated to produce abundant suckers. These suckers rapidly dominate, with slow-growing balsam fir seedlings again developing as an understory, if seed is available. Less than a century is required for the succession from one species to the other because both species are short-lived. In the absence of the other species, either balsam fir or aspen can perpetuate itself with successive generations.

Insects and Forest Succession

Shifts between aspen and balsam fir-spruce domination are often hastened by infestations of spruce budworm and forest tent caterpillar (plates 7 and 8). Severe, early summer infestations of forest tent caterpillar completely defoliate aspen stands for three or four successive years. Trees often releaf later in the summer, but they are weakened and more susceptible to a variety of fatal fungous diseases, hastening decline of the stand. Spruce budworm attacks balsam fir severely, completely stripping trees of needles. Despite the insect's name, it causes less damage to spruce. Three to five consecutive years of defoliation usually kill entire balsam fir stands. The trees fall over in jackstraw fashion, and dead stands become high fire hazards (fig. 6.22).

Although both insects are native to the area, there is no mention of budworm-killed stands in the notes of any explorer or surveyor; prelogging infestations must have been limited in size and frequency. However, some of the impenetrable down timber the surveyors described could have been killed by budworm. Only one mention was made of defoliated aspen. As quoted in chapter 4, a

Figure 6.22. Dead and down balsam fir destroyed by spruce budworm. Photo by authors.

Canadian artist traveling with a Hudson's Bay Company expedition in 1848 noted heavy aspen defoliation along a border lakes route.

Occasionally, fire in budworm-damaged stands reduces the budworm population, and late spring frosts may cause a decline in forest tent caterpillars. Neither reduction is permanent, however, for moths of both species can be carried at least three hundred miles in twelve hours by turbulent air associated with cold fronts.[29] As populations build up in extensive aspen and balsam fir stands adjacent to the BWCA, both insects are quickly reestablished whenever a new forest matures.

Spruce budworm does not damage black spruce seriously, and it is retarded by the presence of white spruce in balsam fir stands.[30] However, the accumulation of debris on the forest floor following a severe budworm attack inhibits regeneration of spruce but not that of balsam fir.[31] Consequently, the repeated insect infestations within the border lakes country may be one reason for the sparse distribution of spruce among balsam fir.

Balsam fir stands older than sixty years are most susceptible to budworm damage, and they are much less susceptible when less than forty years old or when overtopped with nonsusceptible species or mixed with a large proportion of spruce.[32]

Infestations of both insects increased in severity and extent during the past half century.[33] These infestations gained momentum because of the increased acreage of aspen and balsam fir throughout cutover land in northern Minnesota. The insect problems and associated ecological changes resulted from extensive postlogging succession outside the BWCA as well as within the area itself. They must be added to our growing list of human influences in the border lakes forest.

Effects of the Aspen-Balsam-Fir Sequence on the Total Flora

Shrub and herb flora are often ignored in discussions of natural and human influence in the forest. Forest floor litter and duff can, however, inhibit the establishment and growth of some smaller plants, just as it does of tree seedlings.[34] The litter of conifers is especially inhibitory, which is one reason why the number of herb and shrub species growing in conifer stands is much lower than in aspen and birch stands. A walk in the forest with eyes on the forest floor will confirm these differences to even a casual observer. On our permanent ecology study plots, twice as many species were found on the forest floor in aspen stands as were found in jack pine or tall pine stands. The size and weight of such plants is also much lower in conifer stands, the combined result of litter inhibition and reduced light on the forest floor.

The postlogging increase in aspen acreage provided better habitat for the nonnative plants that followed human migration into the wilderness. Thus, many of the nonnative species that constitute 12 percent of the flora became established on cutover land and spread along the edges and within the aspen-dominated forest. The increased aspen forest also permitted increased distribution of many native herbs and shrubs as well.

The ecological effects of early logging, then, are not limited to the removal of tall pines. They also include introduction and spread of disease, increased insect damage, introduction of nonnative plants, and increased acreage and accelerated forest change within the aspen-birch-spruce-fir sequence. Some of these effects originated in cutover lands outside the immediate BWCA. Had the border lakes forests not been logged, the effects of logging in adjacent areas would still have had an impact, but at a much slower rate.

Thus, we can consider our white pine lob tree a sentinel, watching over a forest that has seen many changes triggered by the tall pine harvest. These changes must modify our definition of wilderness to include lands in which human enterprise has severely limited natural reproduction of the two tall pine species, making way for an increase of the aspen-birch-spruce-fir sequence, associated herb and shrub growth, and corresponding changes in wildlife.

Figure 7.1. White spruce lob tree. Photo by authors.

CHAPTER 7

White Spruce Lob Tree

PULPWOOD LOGGING

Our white spruce lob tree stands out as a landmark on the lakeshore (fig. 7.1). Its graceful branches curve gently upward at their tips and are covered with short, sharp needles. The crown tapers skyward, but the tip is fuller than the spiked tips of the smaller balsam fir near-by. A few dry, amber balls of pitch have formed on its trunk. Only hardy souls with strong teeth still chew this spruce gum, usually just to prove they can, but it is nature's multipurpose glue, valued by Ojibway in the construction of birch bark canoes and containers. Black spruce and white spruce, along with the aspen, birch, and balsam fir with which they often grow, are the major tree species remaining in the border lakes forests after tall pine logging. The spruces were the first of these remaining species to stimulate the next phase of logging in the area, so they stand as a symbol of the most recent era of timber harvest.

A New Era of Timber Harvest

In the early 1930s, pine logging had slowed and the big sawmills closed in northeastern Minnesota. Most construction lumber now came from western and southern states. Some logs were left floating in border lakes because there was no market for them. Minnesota sawlog harvest was limited primarily to the production of mining timbers, railroad ties, and lumber for local use. Unemployment in the lumber industry added to the pinch of the Great Depression.

The forests were not idle long, however. Minnesota had a wealth of aspen, birch, spruce, fir, and jack pine lands, some left untouched by tall pine loggers, others regenerated after red and white pine were removed. Wood of these remaining species could be ground to pulp or processed with heat and chemicals and made into paper. Pulpwood cutting began in the central, accessible forests and moved northward, the same path followed by the white pine industry half a century earlier.

Most pulpwood in Minnesota was cut on lands outside the present BWCA. Vast acres of northern Minnesota backwoods were cut. The expanse and forest succession of these cutover lands, and their proximity to the border lakes, indirectly affected wilderness area insect and disease cycles and plant succession, just as the tall pine harvest had.

Pulpwood logging bore some striking similarities to the tall pine harvest in scope, vigor, and completeness of operation. However, changed technology, increased knowledge of forest behavior, a conservation-minded public, and an altered forest mosaic all contributed to different effects on the border lakes forests, part of a slowly evolving influence of human activity in the wilderness. To understand these changes, we must follow the pulpwood industry in its northward progress through Minnesota, just as we followed the tall pine harvest.

Need for pulpwood became urgent, as had the need for pine lumber less than a century earlier. The nation's population was growing, life-styles and learning levels were rising. Business, industry, and governments increased in size and complexity, making communication important. Letters, books, newspapers, and magazines appeared in every home. The paper bag and cardboard carton replaced the market basket, flour sack, and milk bottle. Indoor plumbing created the need for large quantities of a new type of paper. Everything had to be packaged; every package required pulp.

Across the Lake States, paper mills appeared in former sawmill towns. In Wisconsin, river towns of five thousand to seven thousand often supported from three to seven pulp mills. The same rivers that had transported sawlogs now provided hydroelectric power to run pulp mills, water for the pulping process, and handy disposal of mill wastes. Those towns that made the transition from pine sawmill to paper mill before 1930 did not suffer noticeably from the Great Depression.

Unlike pine sawmills, pulp mills did not follow the timber harvest into the wilderness. Papermaking required extensive machinery, water power, and specialized river sites. Pulpwood could now be transported long distances on roads and railroads. One company—

Figure 7.2 Raft of pulpwood assembled at Grand Marais to be floated to Ashland for Wisconsin pulp mills, 1967. Courtesy U.S. Forest Service.

Consolidated Water Power and Paper Company of Wisconsin Rapids (now Consolidated Paper, Inc.)—for a time even floated huge rafts of spruce from nearby Ontario, Isle Royal, and Cook County, Minnesota, across Lake Superior to Ashland (fig. 7.2). The majority of the Lake States paper mills were in Wisconsin, but by 1965 Minnesota had nine paper mills and twelve wood conversion plants that provided employment for 9,000 mill workers and 11,600 woods workers and a total payroll of over $63 million. In addition to supplying Minnesota mills, the forests of Minnesota supplied large quantities of pulpwood to Wisconsin mills.

Although large mills and extensive equipment were required for the actual manufacture of cardboard and paper, pulpwood could be cut by small operators working independently and selling their product to pulp companies or contracting to cut on the company's lands. Many independent settler-pulp cutters sold rail carloads or truckloads of pulpwood from their own or publicly owned land. Small cutting operations were common.

Early pulpwood cutting methods required only simple tools,

Figure 7.3. Pulpwood cutting in black spruce forest of the BWCA. Courtesy
Robert Hagman.

similar to those used during most of the tall pine harvest. The ax,
bow saw, and four-foot crosscut saw were used for cutting; horses
skidded timber from the woods to landings where trucks and rail cars
were loaded. During the late 1940s and 1950s, chain saws and small
tractors began to be used, eventually doubling the daily harvest per
woods worker. Hydraulic lifts speeded loading. By the 1970s, heavy
equipment—harvesters, processors, and chippers—further mechanized logging, even within the eastern and western boundaries of the
BWCA.

Pulpwood cutting reached the periphery of the BWCA about
1940. The long wood fibers of spruce were valuable for the manufacture of high-quality writing paper, and early pulpwood harvest in
the area began with the spruces (fig. 7.3). Hammermill Bond and
Northwest Paper companies were among the first to begin cutting
spruce along the eastern and southern borders of the present BWCA
in the late 1930s.

Cook, St. Louis, and Lake counties, which include the BWCA,

were more than 90 percent forested, ranking among the five most heavily timbered and least populated Minnesota counties. The timber remaining after tall pine harvest was 27 percent jack pine, 23 percent aspen, 19 percent spruce, 14 percent paper birch, 3 percent tall pines, and 14 percent other species.

In 1944, a Forest Service evaluation of Roadless Area timber concluded that "much of the forest is mature or over mature and should be cut to avoid decadence and the inevitable conversion to brush"[1] that would occur in the absence of fire. At that time, the entire Superior National Forest, including the Roadless Area, was managed for multiple use. Under such management, logging was sound timber practice for these short-lived species.

The public was told of the possible fate of these overmature, largely jack pine forests if left uncut. With local opinion behind them, the Forest Service actively sought contracts to harvest the mature stands.

The Tomahawk Sale

The largest pulpwood operation to include BWCA forest was the Tomahawk Sale. Begun on a small scale just south of the current BWCA border in 1941, it was enlarged in 1945 by purchase of a large sale originally made to the Ontonagan Fiber Company of Michigan. The total sale covered 155,000 acres; cutting proceeded for twenty years. The operation was begun by Tomahawk Timber Company to supply jack pine pulpwood to its parent firm, Tomahawk Kraft Paper of Wisconsin. Later, the operation also supplied pulpwood to other Wisconsin mills.

In 1947, Forest Center, a town complete with school, chapel, homes, store, restaurant, and recreational facilities, was built for the several hundred Tomahawk Company woods workers and their families. A well-traveled access road supplied the town near Isabella Lake, which is now within the BWCA boundary. By 1949, a railroad had been built to the Forest Center pulpwood landing (fig. 7.4). The operation also included a sawmill producing aspen lumber.

The entire operation harvested 1,700,000 cords of pulpwood. Paul Bunyan had moved on west in search of sawlogs, but pulpwood logging was big business on lands later annexed to the BWCA between 1940 and 1970.

About 63 percent of the pulpwood cut on the Tomahawk Sale was jack pine, 23 percent spruce, and 14 percent aspen, balsam fir, birch, cedar, and red and white pine. Aspen provided sawlogs or was

Figure 7.4. Forest Center pulpwood landing, Tomahawk Timber Company. Courtesy U.S. Forest Service.

sold for making excelsior, chips, and "mattress wood" shavings. Birch was used for match bolts, boxes, veneer, and railroad ties. Cedar provided poles, fence posts, and mining lagging.

Other Pulpwood Operations

In addition to the Tomahawk Company, several others were cutting on lands now part of the BWCA. These included Northwest Paper, Kimberly-Clark, and Consolidated Water Power and Paper, some of which supplied Wisconsin mills that had formerly used Canadian spruce pulpwood. The names were new to the border lakes, but the economic base as well as some faces and mill sites were not. The Weyerhauser family, which had originally controlled the St. Croix Lumber Company during tall pine logging days, was represented by Northwest Paper Company. Consolidated Water Power and Paper was derived from a Wisconsin Rapids pine logging base, the Witter family. Kimberly-Clark also originated from old Wisconsin pine logging financial bases.[2] At least thirty full-time logging operations had crews cutting on lands now included in the BWCA (fig. 7.5). Most of the logging, however, was done by crews hired and managed by the distant pulp companies.

Figure 7.5. Pulpwood harvest using horses, Tomahawk Timber Company, 1947.
Courtesy U.S. Forest Service.

Much of the pulp land was essentially clear-cut, just as the tall
pine lands had been, except for some mixed conifer sales where only
jack pine was removed. White spruce and black spruce had been the
first species to attract pulpwood cutters, but by the time the harvest
had reached the border lakes forests, some mills were able to use
other species especially jack pine. All major conifer forest species
rejected by tall pine loggers were now of use. Many of the spruce-fir
and jack pine stands were mixed with aspen; a few white pine and
red pine and cedar trees were scattered among them. The access
roads for pulpwood cutting allowed loggers to reach some of the
remaining tall pine stands, and pine logging continued on a small
scale; some aspen, birch, and cedar were also harvested.

Over 262,240 acres, or 24 percent, of the present BWCA was
logged after 1940 (see map, fig. 6.9). Most of this acreage was located in the south central portion. Before cutting, the area was covered with 51 percent jack pine, 48 percent aspen-birch-spruce-fir,

and 1 percent tall pine. The composition was strikingly different from the original vegetation of lands earlier harvested for tall pine, on which 73 percent was tall pine, 10 percent jack pine, and 17 percent aspen-birch-spruce-fir.

The southern location of the pulp-logged portion made it more accessible than most of the interior land that was cutover during the pine logging days. Lack of extensive red and white pine made the land unattractive to tall pine loggers, however, leaving it wild, uncut, and largely untrod before the pulp cutters arrived. Like the eastern and western portions of the BWCA, much of it had burned over frequently before the development of effective fire suppression, as could be surmised from its typically postfire forest species.

As supplies of conifers declined, timber companies converted all or part of their operations for harvest of the one remaining major timber species—aspen. An early indication of this trend became apparent when the Tomahawk Sale cutting was nearing completion. The Tomahawk Company, a subsidiary of National Container Company, moved the woodlands operations back to Wisconsin for utilization of the hardwoods that had regenerated on cutover pine lands. In Minnesota, new and expanding wood conversion, wood pellet, and chipboard factories are fast increasing their use of aspen.

In the Aftermath of Pulpwood Harvest

Cutting was completely banned in the BWCA before extensive aspen harvest reached the border lakes forests. In compensating the timber industry for loss of cutting rights in the wilderness, the Forest Service placed great emphasis on research and development of aspen-based products and industries because aspen is the only extensive forest type left on remaining nonwilderness lands. The vigor of aspen in sprouting and spreading aggressively after cutting and fire assures that another crop of aspen will follow, at least for a time.

Aspen recovery may be temporary, however, since modern whole-tree harvest utilizes the entire aboveground portion of the tree. This process removes up to three times more vital nutrients from the site than would traditional logging practices, which leave branches and leaves to recycle into the soil.[3] Whole-tree harvest of aspen removes more nutrients than does whole-tree removal of conifer species.

On thin, infertile, northern Minnesota soils, long-term management of site nutrients is often neglected in favor of temporary economic gain. Short rotation, whole-tree harvest is exploitation of the site nutrients, which are nonrenewable resources. This removal constitutes a form of mining, or the mechanical equivalent of over-grazing. Serious nutrient depletion will increase as more intensive aspen harvest removes more nutrients. Poor quality stands and brush-land eventually develop after site depletion.

An indication of the possibly grim future for some northern Minnesota forests so managed has been the development of brush biomass studies[4] and a few experimental pilot plants for the utiliza-tion of the total biomass of all woodland fiber, including twigs and brush. With present cutting restrictions, however, this method will not be a problem in the BWCA.

The Forest after Pulpwood Cutting

Forest development after logging on pulpwood lands proceeded much as it had after logging on tall pine lands. The abundant aspen-birch-spruce-fir type reestablished itself readily after cutting. Jack pine was also reestablished on burned lands. Balsam fir appeared on those unburned cutovers on which it had become established as an understory before cutting. Everywhere, aspen sprouted. Foresters working in the area at the time reported aspen predominating through-out the cutover, except where jack pine lands burned and jack pine was reestablished.[5]

Aspen regeneration occurred on some former jack pine lands with light, thin, rocky, soil. This off-site aspen is short-lived, slow growing, and of poor quality. The biomass and canopy of such a forest is so low-grade that foresters consider it one step above brush and of value only as wildlife browse. This conversion to aspen occurred on jack pine sites cut for pulpwood, some yielding as much as forty-five cords per acre.

Tall Pine and Pulpwood Harvests, a Comparison

There were important differences between the earliest pine logging and the more recent pulpwood and miscellaneous logging. First, since cutting was banned in the BWCA before pulp species were exhausted, more land containing these species was never cut than was true for the tall pine. Old jack pine forest is a major portion of the so-called virgin pine of the canoe country today.

Second, between the onset of tall pine logging and the first pulpwood cutting, the area became part of the national forest system, to be managed by professional foresters. The first pine logging had been a freewheeling operation. Even the international boundary was obscure and ignored by pine loggers. Trespass logging occurred on both sides of the international boundary, as for example T68N R13W on the shore of Lac La Croix, a cutover site still classified as uncut. Although the Organic Act of 1897 required removal of cull trees and state law dictated slash burning, these regulations were only loosely enforced because staff was not available and jurisdiction was vague. Management personnel and the accompanying supervision were introduced when the Superior National Forest was established in 1909. Passage of the Weeks Land Acquisition Act in 1911 permitted gradual accumulation of manageable blocks of land in federal ownership.

Logging on federal land was controlled and recorded. Road building, especially for fire control access, increased for a time and then was gradually phased out. Wildfires were controlled wherever possible, and tree planting and timber stand improvement marked the beginning of change in the direction of human activity in the forest, activity that until then had been completely exploitive.

A third difference between tall pine and pulpwood logging can be found in the expectations of the public. Across the country, bitter experience in the wake of tall pine logging had taught that wild lands, like the resources they contained, were not inexhaustible. For a growing number of people such wild lands now had intrinsic value, and they were no longer regarded only as sources of harvestable timber, game, and fish. Conservation forces were acquiring sufficient political power to guide legislation and supervise its enforcement. The Superior National Forest was subject to zoning, supervised cutting, fire control, recreational use, wildlife habitat management, watershed management, land acquisition and exchange, and reforestation.

The growth of forest management in the border lakes country reflects the changing needs of the forest as well as increased knowledge of how to meet those needs. More important, it adds one more facet to human manipulation of the Boundary Waters.

Early Forest Management

Planting and timber stand improvement records and logging boundaries for the Superior National Forest show clearly that the goal of the earliest forest management was protection and improvement of the remaining uncut forests, not restoration of cutover pine land. In the 1920s, it was expected that cutover lands would regenerate naturally, especially when seed trees were left standing, as was then the policy. This was a logical assumption at the time because blister rust had not yet been discovered in the area and the silvics of red pine reproduction was not yet fully understood. Burned-over land, however, was assumed to require reforestation. Much pine cut-over had not yet been acquired by the national forest; the rest was ignored.

Of the more than three thousand acres planted within the present BWCA between 1911 and 1940, 88 percent was located in the uncut portion of the forest and only 12 percent on cutover pine lands. Most planting was done on land disturbed by fire and blowdown. More than 78 percent of the plantings were red pine, the rest primarily white spruce. In addition to the recorded plantings, some foresters filled their pockets with seed before trips into the woods and spot-planted it wherever they believed appropriate.[6] Packets of seed were also distributed with hunting licenses for volunteer hunters to plant as they roamed the woods.[7]

Jack pine, still considered a weed tree, was planted on only four experimental acres. Small white pine plantings were established before the impact of blister rust was realized. The survival of these ranged from zero to less than 1 percent by 1940, another indication that blister rust was already active in the border lakes forest in the early 1930s.

Many of the early red pine plantings along portages, near campsites, and on islands are now often mistaken for natural repro-duction. Site selection and planting methods must have been in tune with the natural forest condition. Examples of these plantings can be seen on islands in Lake Saganaga and Seagull Lake, the Four Mile Portage to Basswood Lake, the former Hubachek holdings, scattered Basswood Lake resort sites, and campsites. Undoubtedly other vigorous, seemingly natural stands of red pine and white spruce, scattered through both the cut and uncut portions of the BWCA originated during this early planting. Confusion of pine plantings with natural pine reproduction is not unusual. Precedence can be found in the occurrence of some European pines, notably *Pinus pinea,* in the Mediterranean region, where plantings are indistinguish-able from natural stands.[8]

Most other timber stand improvement practices in the 1920s and 1930s were also concentrated in the uncut area. Thinning (cutting where the stand was too dense for good growth) and liber-ation (removal of tall, older trees to permit better growth of under-story species) were completed on more than 4,000 acres of uncut forest but on only 375 acres of pine cutover land. Tree pruning and release (removal of low, competing vegetation) were done about equally on cutover and uncut forest, around 300 acres of each.

Management after 1940

Workers and finances for forest management were especially available during the postdepression recovery period of the late 1930s when Works Progress Administration, National Recovery Act (NRA), and Civilian Conservation Corps crews were scattered throughout the forest. Timber stand improvement camps of the NRA on lakes Four and Saganaga, on Hudson, Insula, Clearwater, Jeanette, and Seagull lakes, and on Portage River operated during this time, thinning jack pine stands, liberating young growth, cutting and burning wind-damaged stands, pruning and clearing along portages, seeding, and in one case actually sodding eroded spots. Veteran woods workers tell of more than forty such camps, each with a crew of up to fifty workers, doing this work throughout the Superior National Forest.[9]

In those days, portage clearing and maintenance was rigorously pursued for fire control access, not for recreational use. Fire fighters approached most fires by foot or canoe. Hundreds of miles of telephone lines to fire lookout towers were strung on trees and telephone poles. Although some important fire spotting and transport of fire control crews was done by private planes under contract to the Forest Service, aircraft did not figure prominently in fire control until after World War II.

Supervised tall pine logging continued on a small scale in the 1930s and 1940s in a few places. An excellent example of timber harvest supervised by the Forest Service can be seen in the selectively cut red and white pine stands in the Angleworm Lake area, where the younger red and white pine were left undisturbed. Forty years later, a vigorous stand has developed and is often mistaken for a virgin (uncut) forest by summer visitors who use the recreational nature trails through the stand.

K-V Reforestation

In 1930, Congress passed the Knutson-Vandenberg Act permitting the Forest Service to set aside a portion of timber sale charges for reforestation costs on cutover land. The act had little effect on border lakes reforestation in the 1930s because cutting was very

Figure 7.6. Establishment of a jack pine thinning study for timber stand improvement in BWCA, Lake Four, 1940. Courtesy U.S. Forest Service.

limited. However, it did stimulate establishment and enlargement of federal tree nurseries and the hiring of more trained foresters to manage reforestation and stand improvement work (fig. 7.6).

The K-V funds were available when pulpwood cutting began and sharply accelerated planting and seeding. Unlike earlier reforestation, planting was limited under terms of the act to cutover lands rather than the uncut, burned forest. This practice indicated a better understanding of postcutting forest recovery. Between 1940 and 1960, almost 70 percent of Forest Service planting was on cutover pulpwood lands, 15 percent on cutover old pine lands, and 15 percent on uncut land. Funds for the latter two came from regular budget, not K-V money.

Foresters who supervised use of K-V reforestation funds tell of carefully distributing planting and seeding efforts over those cutover areas in which natural tree reproduction seemed unlikely. Funds made available under the act were only sufficient to replant about one-third of the cutover. Where only aspen sprouted, the land had to be considered adequately stocked and was not planted.[10] As a result, aspen increased strikingly, its invasive root network spreading over many acres where the species was previously represented only by scattered trees or small groves.

Figure 7.7. Red pine plantation on cutover jack pine land, Tomahawk Sale, Polly Lake, 1965. Courtesy U.S. Forest Service.

Figure 7.8. Planting crew in cutover area prepared by disking. Tomahawk Sale, 1947. Courtesy U.S. Forest Service.

Figure 7.9. Cutover near Hudson Lake that was seeded with jack pine from an airplane in 1963. The photograph was taken six years later. Courtesy U.S. Forest Service.

On lands selected for reforestation, red pine, white spruce, and jack pine seedlings were planted (fig. 7.7) or the areas were seeded with jack pine and spruce. Some seeding from airplanes was done. On some sites within the present BWCA, favorable seedbeds were prepared by disking, which breaks up the matted forest floor, or by rock raking, which bulldozes the brush, timber debris, and forest floor into large windrows (fig. 7.8). On Hudson Lake, for example, an extensive area was rock raked in 1962 and now supports a vigorous jack pine forest (fig. 7.9).[11] Competing vegetation was removed by hand from some young plantations. Later, herbicide spraying, sometimes from aircraft, was used.

The Tomahawk Sale area contains the most extensive and best documented reforestation following pulpwood logging in the border lakes country. Among the 132,000 acres of that sale logged within the present BWCA, 38,000 acres or 29 percent was planted with red pine, white spruce, or jack pine or was seeded with jack pine or spruce.

Although funds for reforestation came from pulp company fees under provisions of the K-V Act, the work was done entirely by the Forest Service, except where cutting contracts called for preparation of land for planting by the company. However, pulp companies restocked company-owned lands, cooperated in research, established experimental plantings, and gave financial assistance to other research organizations working in the area. Much of the pulpwood-cutting area had been zoned for multiple use including timber harvest in earlier Roadless Area plans, and still other portions were only recently included in the designated wilderness area.

Our spruce lob tree stands as a reminder that our definition of wilderness must be expanded to include lands recently acquired or managed for timber harvest for more than forty years. It must include lands penetrated by logging roads and rail spurs. It must also include extensive plantations, seeded areas, and lands in which the forest floor was bulldozed, rock raked, and disked to prepare the forest seedbed. It is a land where airplanes passed low, seeding, spraying, and dropping packets of seedlings to remote planting crews. It is a land where people not only harvested timber but also replaced at least part of what was removed, and in other ways restored the forest condition. However, it is a land in which cutting and reforestation have favored the increase of deciduous species in the forest mosaic.

Figure 8.1. Balsam fir lob tree. Photo by authors.

CHAPTER 8

Balsam Fir Lob Tree

RECREATION AND PRESERVATION

In the early morning stillness, the sharp spike of our balsam fir lob tree is reflected in the mirror-smooth surface of the lake (fig. 8.1). Its spiked crown is part of the jagged forest profile so characteristic of the border lakes country, and it contributes to the northern or boreal atmosphere that brings visitors back again and again. Balsam fir, standing thus as a reminder of the lure of the north, is an appropriate lob tree for our consideration of the ecological significance of human recreation in the BWCA.

For practical purposes, our balsam fir would not make a good lob tree because it is too pitchy to be climbed. The smooth bark is covered with pitch blisters. These contain Canada balsam, the resin first used as a permanent mounting medium in microscope slide preparations. Thin bark and abundant flammable resin make balsam extremely vulnerable to forest fire.

Many cones are clustered near the tree's sharp tip. Short, spine-like structures among the cones are the axes left when old cones disintegrated, shedding cone scales, bracts, and seeds. Cones of the spruces, with which balsam fir is sometimes confused, are shed intact.

An allelopathic or inhibitory quality of balsam fir needles and ground debris discourages growth of herbs,[1] thereby limiting the ground flora in balsam fir stands. Lack of light reaching the ground under the dense fir canopy also discourages growth of vegetation. Only a few species of moss and a sprinkling of tiny, four-needled balsam seedlings are found under our lob tree.

Back from the lake in the nearby balsam fir stand, some trees are leaning, others have fallen over, and some have broken off part way up the trunk. The brittle trunk and shallow root system of balsam fir make it susceptible to wind damage, especially when the stand is mature or on a moist site. Spruce budworm also takes a major toll. But the young seedlings are shade tolerant and some survive on the floor of old stands, permitting constant renewal. Stubs, fallen trees, and trees of various ages and heights give the stand a wild, unkempt appearance, further contributing to the isolated, northern atmosphere surrounding our lob tree.

Balsam fir is a major species of the boreal forest. Almost its entire range is glaciated, extending from mid-Minnesota northward to Hudson Bay and the tundra.[2] Studies of the pollen record show that the species was sparsely distributed in early postglacial forests, increasing in more recent times.[3] Balsam fir is favored by cool, somewhat moist or fresh habitat. It can grow in almost any upland soil, often succeeding aspen and birch on disturbed sites, and it is common throughout the BWCA.

Although balsam fir provides cover for some birds and wildlife, as well as a source of browse for moose, the dense balsam fir stand is often deserted. It reminds one of the impressions recorded by Cabot on his trip to the area with Louis Agassiz:

> This stern and northern character is shown in nothing more clearly than in the scarcity of animals in the dense forest. The woods are silent and as if deserted; one may walk for hours without hearing any animal sound, and when he does it is of a wild and lonely character; the cry of a loon, or the Canada jay, the startling rattle of a woodpecker, or the sweet, solemn note of the white-throated sparrow. . . . [It] is like being transported to the early stages of the earth, when the mosses and trees had just begun to cover the primeval rock and the animals as yet ventured timidly forth into the new world.[4]

A New Approach to the Wilderness

Pursuit of this northern solitude (fig. 8.2) and the recreation it provides brought still further interaction between the land, the water, and human life. Recreational priorities have helped shape the destiny of the border lakes forest and have brought a completely new approach—one undreamed of by any of the earlier wilderness users.

A certain recreational potential was an acknowledged part of the area long before creation of the Superior National Forest. In early days, however, the forest was still viewed as a resource to be harvested. The idea of preserving its wild condition was still only the

Figure 8.2. In pursuit of border lakes solitude, 1919. Photo by A. H. Carhart. Courtesy U.S. Forest Service.

misty dream of a few naturalists far removed from the area. No one in those early days could envision the eventual pressures and regulation that would be involved nor the conflicts of interest that would grow among the goals of hunters, fishermen, campers, canoeists, snowmobilers, loggers, and those interested solely in wilderness preservation. Indeed, early canoeists and campers would not have predicted the most explosive conflicts of all—those among various types of recreational users.

Early Recreational Use

Undoubtedly, many people who trapped, fished, hunted, cruised timber, and logged in the border lakes country in early days felt the intangible lure of the land. They relished their evening meals beside a camp fire, were refreshed by a night's sleep on a bed of fragrant balsam boughs, and watched the morning mists drift quietly over the balsam-profiled points and islands. Their enjoyment was a hard-earned by-product of their means of livelihood. Gradually, however, visitors with fishing, camping, and canoeing as their sole intent began to arrive.

There are a few records of those early adventurers. Some were residents of nearby mining and lumber mill towns. A few came from as far away as Minneapolis and Chicago, where accounts of hunting and fishing trips to the border lakes appeared in newspapers in the late 1800s. In 1897, a Philadelphian, W. H. Richardson, and his bride made a honeymoon canoe trip through the area,[5] and early loggers recalled occasional canoeists and fishermen riding free on lumber company railroad spur lines to Hoist and Pipestone bays of Basswood Lake during tall pine logging days.[6] The first real push for recreational camping, fishing, and outdoor life came about the end of World War I when people began to have more leisure time, vacations became part of the way of life, and increased urban living accentuated the need for solitude and a retreat to the woods.

Recreational use was part of the early Forest Service planning. In 1917, F. A. Waugh was commissioned to study recreational use of national forest lands. His report stressed the beauty, enticement, and direct human values of hiking, camping, and sight-seeing in the forest environment. It recommended that such use be given equal consideration to that for forest product harvest in management plans.[7]

In 1919, the Forest Service sent a young landscape architect, Arthur H. Carhart, to the border lakes country to formulate plans for preserving the area's natural beauty and recreational opportunities. Carhart was especially impressed with the network of lakes and streams. He recognized its recreational potential and believed it precluded the need for extensive access roads. The photographs in his report reveal a fascination with the rapids, waterfalls, streams, and rivers connecting the lakes (fig. 8.3).[8]

Carhart's appreciation for the natural water resources provided a valuable base for the first fight to preserve the border lakes from devastation (fig. 8.4). This struggle occurred in the late 1920s when Edward Backus planned to develop an extensive series of industrial

Figure 8.3. Carhart was impressed with the recreational potential of the network of lakes and rivers, as his photograph of the Isabella River near Isabella Lake reveals. Courtesy U.S. Forest Service.

Figure 8.4. Carhart's appreciation of the natural water system in the border lakes provided the basis for the later fight to preserve the area from industrial dams. Photo by A. H. Carhart. Courtesy U.S. Forest Service.

dams that would drastically alter the border lakes and inundate much of the forest. The successful battle to stop his dam building was the first in a long and bitter series precipitated by the conflicting goals of industrial and recreational use. It culminated in passage of the Shipstead-Newton-Noland Act of 1930. The act protected existing water levels, restricted logging near waterways, and ended home-steading on the Superior—all clearly aimed at preservation of the wilderness for human recreational use.

Passage of the Shipstead-Newton-Nolan Act marked only the beginning of a long line of bitter clashes between the conflicting needs of industrial, commercial, and recreational users and the goals of conservationists. There have followed more than forty years of controversy and dispute among individuals, organizations, businesses, and federal agencies. The earlier, small disputes between Indian tribes, rivalries among fur traders, and the push by tall pine loggers for the best timber seem almost trivial in comparison.

In the earlier disputes, the struggle was among users of the same resource. Consequently, the competing parties understood each other. Now, however, conflicts were among various groups seeking different resources or different uses of the area. When their goals were not compatible, as so often was the case, the conflict was fanned by lack of understanding.

Much has been written about the still ongoing border lakes controversy.[9] Many details thus far appear in lengthy court and press records. It is not our aim in this chapter to review skirmishes along the way nor to evaluate or judge motives. We will deal with the conflicts only as they reflect changing human participation in border lakes forest destiny.

Land Acquisition for Forest Management

Throughout this period of controversy, the managing agency—the U.S. Forest Service—was faced with the complex problem of defin-ing boundaries and forming a manageable tract of land. The Forest Service found itself confronted not with a block of national forest to manage but with a patchwork of federally owned land inter-spersed among industrial holdings and private, state, and county lands. Margins were irregular and acreage was difficult to determine.

The original national forest was made up primarily of public domain or unclaimed land that no one wanted. These public lands were often scattered and isolated (fig. 8.5). Much of the land within the boundaries of the present BWCA, however, was not even part of

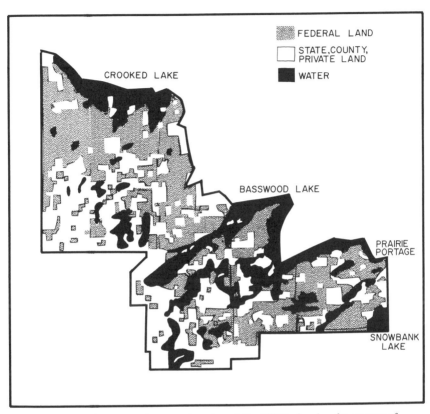

Figure 8.5. Representative central section of the BWCA showing the amount of land in federal ownership in 1937. Area shown is between R8W-R12W and T64N-T66N.

the original national forest. Canoe country land had been bought and sold; fishing camps and summer cabins appeared as the tall pine loggers retreated.

When Congress passed the Weeks Act in 1911, it provided authority for the purchase of additional national forest land. In 1922, Congress passed further legislation that permitted exchange of land among state, county, and federal agencies so that each could consolidate holdings for more efficient management. By exchange and purchase, the Forest Service began the long, arduous, and controversial task of acquiring manageable blocks of land. From 1911 to 1947, land acquisition in the present BWCA under the Weeks Act was solely for practical timber management and watershed protection.

Federal land acquisition was a new concept to the rapidly expanding, frontier-oriented country. The boundary waters country was no longer a free frontier. Between 1936 and 1964, more than 270,000 acres were added to the wilderness area by incorporation of national forest land, exchange, purchase, and donation. Private ownership was reduced from 135,944 acres or 21 percent to 55,645 acres or 9 percent by 1947.

Expanding Recreational Demands

Early management incorporated Gifford Pinchot's concept of managing national forests—the greatest good for the greatest number in the long run. It was assumed that, with proper management, logging and a variety of recreational pursuits could coexist compatibly in the same area. Gradually, an interest in protecting some of the border lakes country from future cutting developed among conservationists. By 1941, a 362,000-acre interior zone was designated as "no cut" to preserve much of the remaining uncut forest in prime canoe country. At the same time, pulpwood harvest and other logging in peripheral zones, especially in the south central section, expanded rapidly. The maintenance of an interior no-cut zone and an outer, multiple-use portal zone—where both logging and recreation were permitted—continued, although not without dispute, until passage of the BWCA Wilderness Bill in 1978 banned all logging in the BWCA.

For many years, the interior, no-cut zone had been open to development. Homesteading had been permitted until 1930. There were also many nonfederal holdings. The original Carhart plan encouraged the building of private cabins. In an article in *Good Housekeeping Magazine* in 1920, Carhart wrote that Uncle Sam would almost give you the ground if you would build a cottage on it.[10] A cabin on a lake had become the dream of many; private lots were sold or leased on lakes made familiar and accessible during tall pine logging days. Resorts were also developed, and between 1941 and 1948 commercial recreational facilities expanded rapidly. By 1950, there were nearly 50 resorts with over 300 buildings on sixteen lakes and more than 150 private cabins on thirty lakes (fig. 8.6).

No estimate of the actual number of people who visited these resorts or used the cabins is available, but traffic by launch, boat, airplane, and amphibious craft was heavy. Buses and trucks maintained regular morning and afternoon schedules over the old railroad bed of the Four Mile Portage. Daily excursion tours of Basswood Lake were conducted from Ely for several years. Some other portages

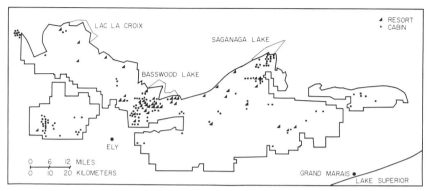

Figure 8.6. Locations of resorts and cabins in the BWCA, 1930-1965. Data taken from Superior National Forest land acquisition records.

were served by trucks and jeeps. The resorts served as takeoff bases for canoe trips, providing motorboat tows to remote ends of the lakes and a hot meal and sauna at the trip's end.

Locations of these resorts and cabins were largely influenced by access, good fishing, and land availability. Many resorts and cabins were concentrated in the former tall pine logging area. There is otherwise no apparent relationship between resort and cabin locations and the sites of earlier trading posts and settlers' cabins. Modern developments were built in a variety of forest types. Many appeared on former timber holdings, especially on shorelines made accessible by logging, on major canoe routes, and near the end of the Gunflint Trail.

Land Acquisition for Recreation and Wilderness Preservation

Banning logging in the interior no-cut zone did not halt the decline in wilderness quality. Cabins, expanding resorts, and the heavy use of associated transportation equipment were not compatible with the solitude and isolation sought by many who wanted to experience the wilderness of the early voyageurs. Therefore, the controversy shifted to wilderness preservationists and recreational users with different goals. Previous uses—logging and fur trade—had been transitory exploitations. The resorts and cabins were more permanent in intent. Pressure to eliminate these permanent dwellings and commercial enterprises grew, but the problem was complex.

The Izaak Walton League was the first conservation organization to take an active role in directing canoe country destiny. An

IWL Endowment Fund for the purchase of border lakes land was set up with initial contributions from clients of the Chicago law firm of Hubachek and Kelly. With these and other contributions, the IWL Endowment Fund purchased land, reserved it from use, and eventually donated or sold it to the Forest Service. Between 1945 and the mid-1960s, the endowment fund purchased and donated fifty-seven tracts of land, or a total of six thousand acres that included a dozen resorts and a number of cabins—no small task for a group of conservation-minded outdoorsmen. In addition, by the early 1970s, F. B. Huba-chek, one of the fund's originators, personally acquired more than twenty-seven hundred acres for donation to the fund or to the Forest Service. With the gift of these holdings, he became the largest individual donor of land to the national forest system up to that time.

Under pressure from the IWL and others, Congress passed the Thye-Blatnick Act in 1948, providing funds and authority for the acquisition of resorts, cabins, and private land in the BWCA. Additional funds were authorized in 1956 and 1961. The IWL Endowment Fund continued to work closely with the government, buying and holding tracts suggested by the Forest Service at times when federal acquisition funds were low.

The goal of land acquisition had now shifted from better management of the actual forest resource to protection of wilderness quality. This new goal was still another human influence in the border lakes country. Earlier acquisition under the Weeks Act had been primarily to obtain consolidated blocks of land as necessary for regulation of stream flow and timber production. Land acquisition under the Thye-Blatnick Act, in contrast, was for removing permanent dwellings, halting commercial development, and reestablishing a primitive atmosphere.

Although much of the privately held land was obtained during the early period under the Weeks Act, acquisition of the remaining land was a longer, more expensive and tedious process that was fraught with conflict (fig. 8.7). This long process is ample evidence of the extent of prior recreational ownership and use of this wilderness.

Between 1948 and 1980, federal ownership grew from 66 percent of the BWCA to 88 percent, while private holdings decreased from 16 percent to less than 1 percent and were eventually eliminated. State and county ownership did not change appreciably, remaining about 11 and 1.7 percent, respectively.

Ownership of BWCA acreage changed gradually during the most active period of land acquisition covered by the Thye-Blatnick Act (table 1). Acreage figures are the best Forest Service estimates

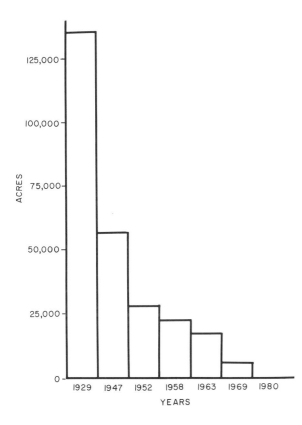

Figure 8.7. Changes in acreage of private ownership in BWCA, 1929-1980.

Table 1. Ownership of Land within the BWCA for Selected Years, Approximate Acres

Year	Federal	State	County	Private	Total
1947	485,108	88,725		55,645	629,478
1949	486,867	97,576		40,759	623,682[a]
1952	496,322	89,098	10,214	26,969	622,603
1955	502,972	87,639	9,937	24,654	625,202
1958	700,608	108,932	14,755	20,935	845,229
1963	736,204	104,111	14,728	17,197	872,240
1979	792,510	106,360	15,420	6,000	921,624[b]

Figures obtained from Superior National Forest land acquisition records.

[a]Early decreases in total acreage result from shifting BWCA or interior and portal zone boundaries. Incomplete records and changes in record keeping are responsible for irregularities in categories and years reported.

[b]Plus 162,480 acres of water surface, making a total of 1,084,104 acres.

available. During this period, acreages in federal ownership were changing rapidly, the outer boudaries of the BWCA were being enlarged, and a variety of acreage figures can be found in the litera- ture. Some do not take into consideration the approximately 21 per- cent of the surface covered with water and under state jurisdiction at the time.

Counties containing the Superior National Forest, and especial- ly the BWCA, are reimbursed by the federal government for federal- ly held land through three different funds. Since its beginning, the Superior National Forest has paid the State of Minnesota one-fourth of its receipts for timber sales. These funds are distributed for support of roads and schools in the counties involved. During the pine and pulpwood logging days, funds from BWCA area timber cutting thereby assisted in the settling of these counties. In addition, since 1976, the counties have received payments in lieu of taxes for federal lands that were on the tax rolls when the federal government acquired them. Again, the many acres of once privately owned land in the BWCA are contributing to county development. Finally, special payments are made to the counties in which the BWCA is located. These payments are based on 0.75 percent of assessed land value. In 1981, for example, payments to Lake, St. Louis, and Cook counties totaled $1,625,709.

Resort and Cabin Impact

Between 1930 and 1970, the major period of resorts and cabins in the border lakes country, most human-caused changes in forest cover were those initiated earlier during the logging era. Most resorts and cabins were built on the shores of large lakes and used only a small portion of shoreline land; users did not venture far inland from the lakes they came to enjoy.

Some additions to the flora, the result of tourist traffic, were discussed in chapter 2. There was some disruption of nesting habitat for waterfowl, shorebirds, and raptors on lakes with cabins and resorts. Much good nesting habitat remained, however, since these lakes with shoreline habitations constituted only 2 percent of the total number of lakes exceeding five acres in size. Wolf and moose populations declined, but wildlife experts associate this phenomenon more directly with changes in the forest after logging and with an epidemic of brainworm among the moose. The brainworm spread to moose from the deer population, which had increased dramatically on cutover land. Deer harbor brainworm but suffer no damage from it.

The greatest impact of resorts and cabins was the decrease in solitude. Pressure for removal of these structures came from wilderness users who believed the establishments threatened the wilderness conditions they enjoyed. The other effects on wildlife and flora were not perceived until long after the resorts and cabins had been removed.

Transportation as a Threat to the Wilderness

Controversy extended from land use to the use of air and water when it became clear that increasing mechanized travel into the wilderness area was threatening not only the wilderness atmosphere but also the quality of remote fishing lakes. If the wilderness state was to be preserved, the threatening advance of airplanes, launches, houseboats, and large motorboats had to be checked.

Air Traffic

Bush pilots had been flying into the border lakes country sporadically since the early 1920s. Many local pilots also carried crews to fight remote forest fires and aided in fire spotting, under contract with the Forest Service, before the Superior had its own planes. Early, somewhat risky air travel was otherwise limited to the occasional trapper, fishing party, or prospector.

When more reliable planes became available in the late 1930s, however, the scene changed rapidly. Airplane base camps were set up on remote lakes, with fishermen flown in and guaranteed their limits of fish. When fishing declined, camps were moved to unfished lakes. The planes also carried guests and supplies to resorts and fishing camps on Lac La Croix, Crooked, Gun, Horse, Knife, Agnes, Kekakabic, Thomas, Frazer, Cherokee, Little Saganaga, Basswood, Seagull, and Big Saganaga lakes.

In 1949, as many as twenty-five local planes operated out of Ely. These, along with increased air traffic from Chicago, Duluth, Minneapolis, Cleveland, and St. Louis, made Ely the largest freshwater seaplane base in America at that time.[11] Other planes based in Tower, Orr, Crane Lake, Hibbing, Coleraine, Virginia, Grand Rapids, Eveleth, Bemidji, and Grand Marais flew into the BWCA regularly. In 1944, only eleven planes with a seating capacity of 30 operated in the border lakes country. Four years later, in 1948, the number had grown to sixty-nine planes with a seating capacity of 226.

Most visitors were flown in for serious fishing and were intent on getting their limits. They came, as their ancestors had, to harvest a resource, even though they now did so recreationally. One indication of their impact can be seen in the changes in BWCA use during the period of greatest air traffic increase. In 1945, about 7,000 visitors entered the area. Of these, one-fourth came to fish and hunt; the rest came to camp, canoe, and sightsee. After World War II, use increased rapidly. By 1952, total visitors had increased to 57,000, of whom 77 percent came primarily to fish and hunt.[12] Lack of roads was no longer a barrier to development and use, and the removal of permanent buildings would not stem the flow of recreational sportsmen.

By 1948, conservation groups as well as local people were alarmed by the extent to which air traffic was growing and threatening the quality and potential of the area. In 1949, after hot dispute among plane owners, the Izaak Walton League, and other organizations, President Truman signed an executive order forbidding private and commercial flights over the BWCA below four thousand feet above sea level. The ban was formulated along the lines of military security banning procedures. The contest, defiance, and court litigation that accompanied the ban continued into the mid-1950s. The plane ban was upheld, however, the first in the nation for wilderness protection.

Two other types of mechanized travel increased and came under scrutiny between 1950 and 1970. Each in turn was judged a threat to wilderness quality and was eventually banned or restricted against the protests of those who used it.

Motorboats

Boat motors were banned in certain zones in 1965. Motorboat use has become increasingly restricted since then. In 1979, motorboats were banned completely in the adjoining Quetico Provincial Park. Motorized travel on portages was gradually eliminated except for single franchised carriers on a few heavily used entry portages.

Snowmobiles

In the 1950s, a few resort owners began to use snowmobiles to haul fuel and other supplies over the ice in the winter. These were large, powerful work vehicles in very limited use. However, as smaller, reliable recreational models became available, snowmobile use in the wilderness mushroomed, just as had use of the airplane. The snowmobile suddenly opened the border lakes to winter trapping, fishing, and other recreational use. It made the most remote lakes accessible again, as the airplane had done earlier. This time, these lakes were visited during winter months, a season when they had seldom or never been visited before.

By the 1960s, snowmobiles were parked at the back doors of many homes throughout northern Minnesota. Every weekend, cars pulling snowmobile trailers headed north from Duluth, the Twin Cities, and farther south. People rushed to enjoy the new experience of winter in the remote northern forest. Many used lakes and trails in other parts of the Superior National Forest, but for a few short years, they also roamed the lakes and portages of the BWCA, limited only by rough terrain, thin ice, open water, dense brush, and the amount of fuel they could carry.

Although winter snowmobile use did not conflict with summer recreational use, it was incongruous with the total wilderness condition sought by preservationists and could come into conflict with snowshoe and cross-country ski activity. Its possible effect on area wildlife was also in question. Wilderness preservation groups from all over the country demanded that snowmobile traffic be stopped. In addition, since snowmobiles were not in use before the Wilderness Act was passed, there was no "established use" argument for their continuation. The Superior National Forest supervisor, Harold Andersen, made the decision that their use should be gradually phased out, following procedures for local guidance and participation set up by Gifford Pinchot half a century earlier:

> I met with key people and discussed this problem at a series of meetings. We proposed to gradually terminate the use by 1980 and gained general acceptance of this by snowmobile clubs and most local citizens. The Sierra Club, after the 8th District Court of Appeals reversed Judge Miles Lord, made an Administrative Appeal to the Chief (of the Forest Service) requesting that snowmobiles be terminated immediately. . . .
>
> This decision created a furor among local people. As a result of Congressional delegation pressure, the U.S. Forest Service reversed its decision. These actions created another furor, for all parties were now up in arms. Had the plan been permitted to stand, the snowmobile problem could have terminated in an orderly manner. In this case the Sierra Club's action backfired against them because they could not accept the fact that in some cases a little patience and time can often help resolve difficult problems.[13]

Because of local displeasure, use of snowmobiles was extended four more years; they were banned in 1976, except for a few designated routes.

Although protests against the sudden ban resulted in an extension of wilderness snowmobile use for four years, snowmobilers and their state, local, and national organizations were ultimately no match for the ever-growing power and influence of wilderness preservation groups.

The snowmobile issue presented the first conflict of interests between actual wilderness users and a group including many nonusers who differed in their concept of wilderness. The successful effort to ban or restrict snowmobile use within a short ten-year period after its introduction is dramatic evidence of the growing power of wilderness preservation forces throughout the country. For the first time, the fate of the border lakes country was guided in large measure by a group of people, many of whom had never seen the area, and who sought no resource or recreation for themselves. At the same time, the phenomenal growth of snowmobile use in a five-year period attests to the pressure of merchandising, mechanized travel, and the growing recreational population.

Wilderness as a Crusade

Wilderness preservation per se became a cause to champion. The preservation movement had begun in the late 1800s; organization of the Sierra Club in 1892, first led by John Muir, was a landmark. The efforts of wilderness preservationists were strengthened by a growing public realization that federally held lands are owned by all the citizens. Each citizen, no matter where he lives, has a right to a voice in federal land management.

Gradually, many influential forces joined the efforts to protect and save undeveloped land throughout the country.[14] Their goal is protection of a land condition that they consider important to preserve. Their personal use of the areas for which they fight is secondary or nonexistent, unlike that of earlier wilderness preservationists who fought for a primitive condition primarily because of their desire to use and enjoy it.

For wilderness purists, as the most dedicated of these crusaders are now known, the preservation of land in the undisturbed state goes beyond the good earth management that, as conservationsts, they also endorse. They see wilderness as a hierophany, a sacred manifestation of the absolute: "He can immerse himself in its perfection and emerge purified. This is strong medicine. When the wilderness ethic is seen in its religious or spiritual context, it is easier to understand the emotional heat generated in purist political struggles."[15] The purists have become the most ardent wilderness crusaders. They share:

> . . . a sense of the tyranny of civilization, even though they are
> inheritors of a long tradition of exerting that tyranny over wilderness.
> Collectively, they are highly urbanized, well above the national average
> in education and income, highly sensitive, and capable of exerting
> influence wholly disproportionate to their numbers.[16]

These preservationists and purists, together with an ardent and growing army of campers and canoeists, became an invincible power in shaping wilderness policy and future human influences in the BWCA.

And Still the People Come

Between 1949 and 1980, recreational use of the border lakes mushroomed, regardless of restrictions, decreased private holdings, and a decline in mechanized travel. Total numbers of visitors rose from an estimated 2,000 in 1940 to well over 100,000 in 1980.[17]

To achieve a realistic estimate of population pressure in the area, recreation specialists have determined "visitor days" by multiplying the number of visitors by the length of their stay. Since 1964, a visitor day has been defined as twelve hours per person or the equivalent. Three people staying four hours, for example, would constitute one visitor day, as would other numbers and hours that would make up a twelve-hour period. Before 1964, a visitor or man day was defined as one person staying for twelve hours or any fraction thereof—each visit, however short, would constitute one visitor day. For this reason, early figures are slightly higher for the same amount of use than are the figures since then.

Regardless of calculation methods, visitor days climbed from 38,000 in 1942 to 1,105,208 in 1975 (fig. 8.8). In the period 1966-1969, recreational visitation for the BWCA was greater than in any other wilderness area in the country. The early restrictions on motorized travel and the closing of resorts and cabins slowed the increase only temporarily, as in 1952-1955 during the airplane ban controversy and 1960-1965 when the major resorts were closing.

As more visitors came, they stayed longer. The average stay ranged between 2 and 4.5 days from 1943 to 1958, typical of the weekend and long-weekend fisherman. From 1966 to 1980, the average stay fluctuated between 6 and 7 days.

Visitor goals and travel patterns also changed. As we have seen,

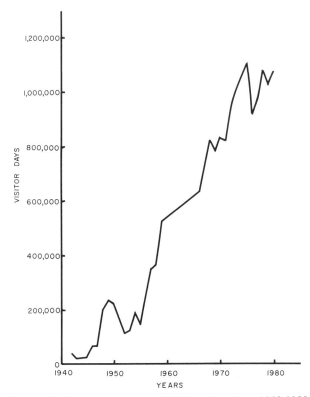

Figure 8.8. Changes in total annual BWCA visitor days, 1940-1980.

fishing parties became more numerous in the BWCA in the late 1940s, stimulated by easy airplane access to isolated lakes, motorboat travel, and fishing camps. During this period, fishing was the major purpose of 75 percent of the visitors. The purpose of recent visits is not directly available in BWCA registration data, but it is reflected in the changing trends in travel mode. Some canoeists may fish, but a greater proportion of visitors in motor-propelled craft come primarily to fish. Paddle canoes accommodated 52 percent of the 1972 visitors, while motorized canoes and motorboats brought 48 percent (fig. 8.9). By 1980, 74.4 percent of the visits were by canoe and only 25.6 percent by motorized craft. Therefore, we may assume that many more recent visitors came for the wilderness experience with only secondary or no interest in the harvest of a tangible resource—fish.

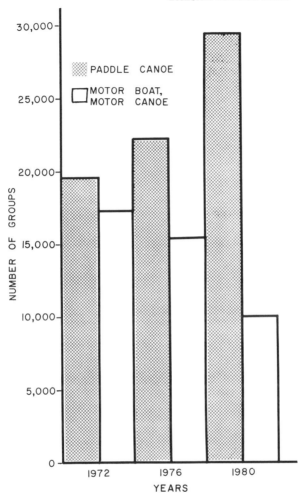

Figure 8.9. Recent BWCA trends in mode of travel.

Gateway to the Quetico

The BWCA entry points have for years also served as gateways to Quetico Provincial Park for those who sought a longer, more remote camping or fishing experience. In 1972, 16 percent of the BWCA visitor groups went on into Canada. With increased Canadian restrictions and fees, travel to Canada dropped slightly to 12 percent of the parties in 1978. Significantly, 74 percent of the 1978 Canada-bound parties paddled canoes, in contrast to only 53 percent in

Figure 8.10. Many come in search of solitude and the pristine quality of border lakes, as in this view of Seagull Lake. Courtesy U.S. Forest Service.

1972, a change in proportion similar to that for total BWCA traffic. Motorboat traffic was banned completely in the Quetico in 1980, after several years of motor horsepower limitations.

Preservation, Publicity, and People

It is ironic that the sharpest increases in visitor use followed the greatest efforts to protect the wilderness quality of the area: 1955-1960, after the airplane ban was upheld; 1965-1968, after passage of the 1964 Wilderness Act; and 1971-1975, when snowmobile, logging, and mining restrictions and ban-the-can regulations were receiving heavy national media attention. Because the conflicting interests were newsworthy, they increased awareness and stimulated interest. Could the very efforts made to preserve the wilderness quality constitute a form of advertising that brings a deluge of visitors?

Those who came in search of wilderness solitude and pristine quality were, by their very presence and wear and tear on the area, diminishing the solitude and quality for other modern voyageurs (fig. 8.10). Although their use of the area is termed nonconsumptive resource utilization, it does leave definite marks on the wilderness through campsite erosion, impact on the trails, litter, and pollution. Although this use may not constitute trammel in the wilderness sense, it does provide "trample" and the destruction of solitude and pristine quality. The paradox was recognized as early as 1949 by Aldo Leopold in other wilderness areas: "All conservation of wildness is self defeating, for to cherish, we must see and fondle, and when enough have seen and fondled, there will be no wildness left to cherish."[18]

Extensive border lakes recreational use in the 1960s and 1970s was the product of a growing population, increased urbanization, increased awareness of wilderness values, and publicity about BWCA controversies. This use demonstrated that even without harvesting a resource, human presence in remote areas can have a marked effect on wilderness quality. Litter, use of forest fuel, species introduction, destruction of wildlife, vandalism of trees and historic sites (especially by improperly supervised youth groups), campsite erosion, change in forest fire patterns, use of undeveloped campsites, and sanitation problems all were taking their toll.

This deterioration had happened elsewhere. In California, the Sierra Club had pressed for controlled access to certain mountain meadows because of the threat of overuse as early as the mid-1940s; it had been concerned about overuse of the Yosemite Valley in the 1930s.[19]

Control of canoe traffic in the border lakes developed gradually. It first became necessary to restrict camping to designated campsites, closing those sites that had become badly eroded from overuse. Latrines and fireplace grills were placed on sites along regularly traveled routes because many sites were occupied every night during the peak season. During heaviest use in July and August, finding a vacant campsite in midafternoon or early evening on well-traveled routes near access points was a lucky event.

The Forest Service increased designated campsites from 1,667 (accommodating 6,520 people) in 1969 to 2,300 (accommodating 11,500 people) in 1974. Closing of overused campsites dropped the number to 2,200 (12,000 people) in 1980 (fig. 8.11).[20] The closing of small, one-canoe campsites on islands and the establishment of new sites accommodating three canoes resulted in greater camping capacity. Some of these sites are located on popular routes and are heavily used; others, on more remote lakes, are not used as frequently. Although major movement still follows old voyageur routes, canoe parties are encouraged to move into less used travel routes. The general picture illustrated in figure 8.11 is the planned dispersal

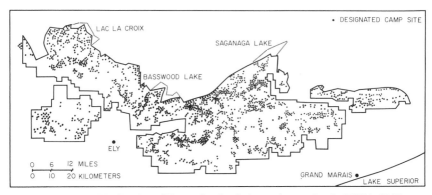

Figure 8.11. Distribution of Forest Service designated campsites in BWCA, 1982.

of canoe parties throughout the border lakes. Increased traffic has widened human wilderness contact to an extent never approached in previous wilderness uses. Additionally, many one-day visitors and fishermen who do not spend the night at a campsite still use the area.

Increased restrictions, increased controversy, and increased use went hand in hand. During 1975—the year of heaviest use, with more than 1,090,000 visitors to the area—there were probably more people in the border lakes country than had been there during all the early years of Indian life, exploration, fur trade, and logging combined. Had the seasonal migration of any other nonnative, large mammal into the area burgeoned so vigorously, it would have been cause for alarm, for at least an open hunting season, and possibly a bounty! Therefore, it is not surprising that limitations on this large and increasing number of wilderness visitors became necessary.

People Management

Despite camper education, vandalism increased. Young trees and saplings on campsites were cut (figs. 8.12 and 8.13). Campsites eroded, especially on islands and at the pine campsites preferred by campers.[21] Water quality near campsites deteriorated; coliform bacterial populations, phosphate concentrations, and turbidity increased.[22] Despite antilitter posters and "pack-it-out" litter bags, refuse problems also increased. Latrines and lake bottoms were accumulating cans and broken glass at increasing rates; plastic container fragments became part of the campsite forest floor. Portages and campsites on heavily used routes required almost daily maintenance work.

Figure 8.12. Red pine thoughtlessly cut down by campers, North Bay, Basswood Lake. Photo by authors.

Figure 8.13. Use of red pine branches for lean-to on North Bay campsite. Photo by authors.

Figure 8.14. Canoeists had always explored freely wherever the lakes, maps, and portages took them. Courtesy U.S. Forest Service.

Pressure for ban-the-can regulations mounted. The Forest Service, under siege for the enforcement of increased transportation restrictions, interpretation of the 1964 Wilderness Act, and other problems, was reluctant to impose further restrictions at that time. However, pressure from local canoe outfitters who had first suggested the regulation forced the issue. In this case, increased restriction of freedom was actually sought by users themselves. Federal regulations banning use of cans and bottles in the BWCA were welcomed by many canoeists when they went into effect in May 1971.

With increased use, "mass recreation was possible; mass solitude was not."[23] Traditionally, canoeists had always explored freely, wherever the lakes, portages, and maps took them (fig. 8.14). Freedom to roam at will had always been an accepted part of canoe travel. But this freedom was no longer possible. Further restrictions were imposed. Canoe party size was limited to ten to prevent campsites from being used by twenty or more campers at one time. By the late 1970s, use of the BWCA required advance reservation; by 1982, a reservation fee was required. Load limits were place on travel zones; visitors had to choose zones with unfilled limits. Increased restrictions, fees, and fewer entry points were also imposed in Quetico Provincial Park.

The efforts to preserve the elusive quality of the wilderness experience and its even more elusive biological state are presenting challenges far greater and more difficult than the challenges faced by those who originally sought to tame the border lakes country and harvest its resources. In meeting these challenges, we must recognize and acknowledge prior human recreational use. Our definition of wilderness must include this past trammel and the continued heavy use of the area.

Figure 9.1. Aspen lob tree. Photo by authors.

CHAPTER 9

Aspen Lob Tree

CONCLUSIONS

It is just a little thing, our aspen lob tree—inconspicuous, thin, spindly, and only between six and seven feet tall. It is indistinguishable from the thousands of others around it on this recently disturbed site (fig. 9.1). Although it is small, our lob tree and the surrounding members of its clone have achieved their present height in less than four years, making them the most rapidly growing trees in the forest. The number of these aspen suckers in the BWCA, their rapid growth rate, and the ecological forces that stimulated their production are the reasons for selecting an aspen sapling as lob tree for our final chapter on the total impact and significance of human life in the BWCA.

Aspen, with its vigorous sprouting from an invasive root network, is ecologically resilient. A tree species must have this resilience to survive and increase in a forest periodically destroyed by natural and human forces. As we have seen, inability to adapt and reproduce vigorously amid changing disturbance patterns has led to the gradual decline of tall pine species in both the primeval and postsettlement forests.

Aspen is short-lived and therefore is often considered a temporary, postdisturbance species. Once it becomes established in an area, however, its root system spreads and survives even when other species dominate the forest canopy. When the forest is again disturbed, aspen root networks will produce many sprouts and cover an area even larger than did the previous aspen stand. The root network will then invade farther, and the increase will continue with each disturbance. Disturbances of human origin have accelerated this

171

aspen invasion. Aspen, therefore, has truly followed human footprints through the wilderness.

In preceding chapters, we have pointed out human footprints
around each of our ecological lob trees by describing evidence of past
and present human influences on the forest. It is important to the
survival of the forest that we recognize those footprints and their
significance, for we continue to make them. Determination of their
effects is the first step in development of sound methods of maintaining the wilderness forest. Let us retrace those footprints quickly,
reclimb those lob trees, and see what now lies spread out before
us.

The Presettlement Forest

The postglacial, presettlement forests developed through slow,
irregular changes from barren rock and gravel to tundra, boreal
forest, some pine forests, and gradually increasing diversity. In more
recent times there has been a slow decrease in pines and an increase
in boreal and deciduous species. With each disturbance the forest
regained elements of its past, but the proportion of each species in
the forest mosaic never quite returned to its previous state. Further
sensitive, progressive change, then, is what we could expect to find
following human footprints through the border lakes country.

The Wilderness Flora

Human activity has increased the total number of plant species in
the BWCA by 12 percent. Only a small portion of this increase has
been intentional, for the area is not conducive to cultivation. Most
intentional introductions are horticultural species, flowers and
ornamental shrubs planted around resorts and cabins. Very few of
these have naturalized and spread. Weeds and grasses, however, were
introduced unintentionally in feed for horses and oxen during logging and settlement days and by the boots, pants cuffs, and packsacks of canoeists and fishermen since then. Still other plants are
disseminated by natural agents: wind, birds, and water from nearby
inhabited areas to the south and east. These plants might have
entered the BWCA without actual human presence there. However,
their establishment nearby was caused by settlement, and their
introduction into the border lakes country is therefore indirectly of
human origin.

Most invading species require disturbed land, exposed soil, and lack of competition. Both natural and human-caused disturbances have fulfilled these needs.

A few shade-tolerant plants, like goutweed, have invasive and allelopathic potential to dominate sites and impose restrictions (trammel) on the native flora. Weeds may also serve as hosts for exotic viruses, insects, and fungi that could seriously damage the native flora just as the introduced blister rust did. These possibilities have not been fully investigated.

Since nonnative plants have not altered the forest mosaic dramatically, their trammel is probably minimal. Thus far, they contribute to a more varied flora and serve as footprints, interesting reminders of the area's rich human history. A sensitive canoeist can see the blueberry fields on United States Point, the lilacs on Wenstrom's Point, the sweet William on the shores of Lake Saganaga, and the occasional forget-me-not along the portages as adding another dimension to his understanding of the border lakes without really endangering his solitude or destroying his wilderness experience.

Forest Fire

Fire has influenced the age, distribution, and species complex of the forest since early postglacial times. Prior to settlement, the frequency and size of fires were determined largely by changing climatic conditions. During droughts, fires were frequent; the forest seldom grew undisturbed for more than a century. During cool, moist times, fires were fewer and smaller; many parts of the forest grew larger and older.

The influence of Indians, voyageurs, and early settlers on the total fire pattern was relatively small. With the advent of logging in the mid-1890s, however, human influence on fire patterns became more evident and continues with more recent increases in recreational use. In the 1970s, for example, fires of human origin accounted for over three-fourths of the acreage burned.

Fires of human origin occur in spring and fall as well as during summer months, in contrast to natural, lightning-caused fires that typically occur in midsummer and late summer. This human extension of the fire season brings concomitant changes in the influence of fire on vegetation. Spring fires favor vegetative sprouting of deciduous species more than do summer and autumn fires. Since fires of human origin cause the greatest increase in burned-over acreage in the spring, this creates a postfire trend toward increased deciduous recovery, especially of aspen, birch, and brush.

Natural, lightning fires occur primarily during drought years. While human-caused fires increase in dry years, they also burn substantial acreage during years with normal rainfall. Their addition to the fire pattern contributes a more varied age to the postfire patches in the forest mosaic.

Fire suppression efforts, beginning in 1910, limited the size of many potentially large fires. Some large fires do still occur, however, under conditions of high fire hazard. Fire prevention, through increased public awareness, improved camping practices, and burning regulations, also contributes to the decrease in incidence of human-caused fires from about 80 percent of total wildfires in the 1940s to slightly more than 50 percent in the 1970s, although the area burned by fires of human origin is still large.

The extent to which fire suppression and prevention have reduced the impact of fire needs further study. Some ecologists believe that fire has been almost eliminated and must be reintro-duced into the forest if preexisting sequences of natural succession are to be maintained. Fire control specialists disagree and believe that, at best, fire suppression barely offsets the increase in fires caused by human activity. Certainly the Superior National Forest and Quetico Provincial Park fire records for the past four decades support the latter view, as does the steady increase in jack pine and aspen postfire types.

Increased fires of recent years may also be related to a gradual buildup of fuel in older forests during the past seventy years of fire suppression. Deciduous species act as fire barriers, however, and increased acreage of aspen can limit the size of fires in some places. Should aspen acreage continue to increase, it could have a significant effect on total fire incidence and resultant vegetation patterns, all ultimately the result of human trammel.

In the past century, human activity has altered fire season, fire distribution, fire years, and acreage burned. Coupled with the re-moval of tall pines during early logging days, it has altered the ability of the forest to respond to fire and has changed fire risk, fire hazard, and damage. These human footprints are evident in the present BWCA vegetation and must be recognized and carefully considered if the forest is to be maintained in its wilderness state.

A change in the response of the forest to fire or other distur-bances can influence even aquatic life in subtle ways. For example, smallmouth bass fry feed voraciously during their first week after hatching on tiny water fleas that always appear in great numbers, stippling the calm shallow water just when the bass hatch. Under the microscope, we can see that the digestive tracts of the water fleas are full of fragments of pine pollen that is shed abundantly at that time and that sifts down to coat the water with a thin, yellow film. The bass fry could probably find a food source other than the water fleas, and the water fleas could find food other than pine pollen. However, this energy relationship exists now, and without the pines it would not.

Human Influences before Logging

Human use of the border lakes forest in the prelogging era was too light to leave extensive footprints. However, those early inhabitants used the resources just as did many of the people who followed them. This view of the wilderness as a resource to be harvested provides interesting comparisons with more recent "purist" attitudes.

The resident Indian population was never large—far smaller, in fact, than one typical summer's roster of canoeing Boy Scouts. As residents, however, the Indians adapted their life-style to the available forest resources. In the absence of European influences, they harvested what they needed on a day-to-day basis altered only by frugal drying and smoking of food for winter and travel use. What they consumed was replaced by nature's bounty with little or no effect on the total forest. Had their numbers increased—say tenfold —however, their impact would probably have been noticeable. The extent to which their frugal life-style was based on a sound feeling for resource conservation is questionable.

The fur trade brought the first serious changes in human resource harvest. Fur-bearing animals were trapped in far greater numbers than required to clothe the resident population. The impact of fur trapping was largely temporary and probably did not greatly alter forest cover. However, it marked the beginning of harvest of border lakes forest resources for profit.

The number of European settlers entering the area in the 1800s was even smaller than the resident Ojibway population, and its impact on the forest was also slight. Unlike the Indians, however, settlers attempted to alter the forest, to create nonnative agricultural products. They failed; the country's soil and climate defied cultivation. Had farming been profitable, much of the forest would have been removed (trammeled), methodically and permanently. Settlers found more hospitable soil to the south and west, and the surge of settlement outside the area placed the border lakes forests near centers of population that would later affect their use.

For a time, miners and prospectors roamed the lake region, establishing test pits and a few mines. But the resources the miners sought were not there. Had a profitable gold or silver lode been discovered, the mining industry and associated population would also have tamed (trammeled) the wilderness. Had the richest iron deposits been centrally located in the border lakes country, the area would have become the site of Minnesota's iron mining industry. Thus the glacier, with infinite wisdom, set down far-reaching limits on the direction of human use, and it still sets limits today.

Timber Harvest

Like fur traders, early loggers found the resource they were looking for, took it, and moved on. The forest itself became the resource to be harvested. Logging brought people and industry. More important, it removed vast quantities of red and white pine and became the first human enterprise to leave extensive, permanent footprints that determined future forest conditions.

Red pine has exacting requirements for natural reproduction. Logging destroyed the large supply of seed that was necessary for red pine regeneration. In so doing, logging precluded the natural return of red pine except in limited quantities. The more adaptable white pine was dealt a serious setback by the introduction of white pine blister rust, which now virtually eliminates natural white pine seedlings in the BWCA. Human activity has thus destroyed the natural resilience of these two major species. Aspen, birch, fir, spruce, jack pine, and brush are replacing them.

The replacement is not absolute; some pine stands and scattered old pine trees still remain. Young reproduction is sparse, however, and young, natural tall pine stands are absent. A future tall pine forest cannot return without further human intervention, and its absence is the direct result of logging (trammel).

Pulp cutting followed on the heels of the tall pine lumber industry, using those species the pine loggers had rejected. Cutting was stopped in the BWCA before the supply was exhausted. Pulp species are more resilient than the tall pines: aspen sprouts vigorously after disturbances; jack pine and black spruce both seed in well after fire; white spruce and balsam fir regenerate under jack pine or aspen in some places.

The type of forest that is established after logging and fire, individually or in combination, is determined by the efficiency with which the available species can reproduce in the altered environment.

Birch, aspen, and some brush species, while reproducing from seed on some sites, also reproduce vegetatively after disturbances and thereby have an extra advantage in their early start, rapid multiplication, and vigorous height growth. Consequently these species tend to become rather evenly distributed throughout the forest. In contrast, conifers lack this vegetative resilience and have, in their dependence on seed reproduction, a weak link in their continuity. As a result, conifer distribution tends to become patchy and less uniform. This discontinuity increases in the cutover forest and will continue to do so.

Forest Management

Establishment of the Superior National Forest added management practices to the list of human impacts on the BWCA. Most early work was related to fire suppression, but it also included tree planting, release, and thinning. Many stands planted during those early days are now indistinguishable from natural forest. In a 1921 newspaper interview, the Superior National Forest supervisor commented that "areas that have been cutover for timber are expected to be restocked by natural forest, while burned-over territory will be reforested artificially."[1]

Some present-day ecologists believe that the reverse is true: burned lands regenerate, cutover lands do not. Both approaches overgeneralize, with one correct on some sites and the other correct on others. Careful evaluation of each site is required.

Forest management accelerated during pulpwood cutting with the availability of K-V funds. Since these funds provided for reforestation of only about one-third of the cutover, many acres covered with aspen sprouts were considered stocked and were not replanted with conifers. Here, as in miles of northern Minnesota outside the BWCA, aspen increased strikingly.

Aspen, the Ignored Invader

Aspen is often overlooked in forest descriptions and in considerations of forest type, and this neglect has several reasons. First, many ecologists classify the border lakes aspen forest as spruce-fir forest type in anticipation of the ultimate domination of the young fir and spruce seedlings they find growing on the floor of the aspen forest. It is true that aspen is short-lived and often occurs as a temporary overstory, gradually replaced by balsam fir and spruce. Even when some aspen stands are less than ten feet tall, a thick sprinkling of balsam fir seedlings can be found beneath them. Many such seedlings die each year, but others replace them, and during the life of an aspen stand, a new balsam fir forest can develop beneath it. Balsam fir, however, is also temporary and short-lived. In its decline, aspen resprouts and the aspen forest is renewed. Aspen can adapt to a wider range of conditions than can balsam fir, so with each aspen cycle, its territory is enlarged. Since conifers have been prominent in the past, the tendency is to assume a continuing alternation of these species. A possible progressive, gradual increase of aspen in the forest mosaic is often ignored.

Secondly, we tend to see what is of value or interest to us. For example, early explorers largely ignored the forests in their journal notes, concentrating on waterways, routes, and fur-bearing animals. Their notes on the forest are sporadic and not comprehensive. Numerous contradictory statements can be found. In 1895, U. S. Grant, field geologist on a survey, noted luxuriant growth of pine, spruce, balsam fir, and birch on the border route between Lake Superior and Rainy Lake. He does not even mention aspen.[2] Only eight years later on the same route, E. A. Braniff, evaluating the government's forest reserves, noted that the land had once been covered with conifers but that these had now been replaced by dense forests of aspen and birch. Just as jack pine was considered a weed and ignored by tall pine loggers, early pulpwood companies often ignored aspen because their mills accepted only conifer species. In the deciduous forest, especially in the eastern part of the border lakes country, birch was prevalent; since it was more valuable than aspen, the forests were classified as birch-aspen.

Finally, the location of the BWCA contributes to a confusing interpretation of its forest succession. Situated roughly parallel to the western extension of a line along the north shore of Lake Superior, the BWCA lies in the transition zone between the true northern or boreal forest and the northern pine and deciduous types. In the boreal forest, spruce and fir, once established, usually maintain themselves in successive generations. Aspen, although a component of the boreal forest, is scattered and limited. In the transition zone forest of the BWCA, aspen is more extensive.

As a result of its resilience, aspen has invaded land not completely suited to its optimum growth. Such off-site stands are often slow growing and of poor quality, lowering the health of the forest in terms of biomass production and canopy. Even on such sites aspen resprouts abundantly, its root network invades, and the species is perpetuated.

Before settlement, a progressive change to a different forest type occurred slowly over thousands of years, guided only by climate, natural disturbance, and changes in habitat created by the species themselves. Human activities, especially timber cutting, clearing, and alteration of the fire pattern, have dramatically accelerated the process by favoring vegetatively reproducing species such as aspen. The extensive aspen forests of the border lakes are, then, at least in part, human footprints in the wilderness. It has been estimated that only about 7 percent of Minnesota's aspen forests will convert to balsam fir-spruce in the next rotation.[3] The percentage is higher in the BWCA, but aspen stands lacking potential to convert to conifers are common.

Recreation

Recreational use of the border lakes country was low during the pine logging period and increased slowly under multiple-use management. Few could envision its phenomenal growth and impact. In 1975, 1,090,100 visitor days were logged in the BWCA, making it the most heavily used wilderness area in the national forest system, with a yearly average of about one visitor day per acre. Since most of the million-plus acres are never actually touched by human feet, this puts tremendous pressure on shorelines, islands, trails, and campsites along well-traveled routes. Water travel acts as a buffer, of course: were the miles traveled by canoes and boats covered on land, the sheer trampling would become trammel.

Recreational use added the concept of wilderness as a place for play and enjoyment. While early recreational visitors were primarily fishermen, their harvest was not for economic gain, and fishing as a primary goal declined over the years. Thus, except for service enterprises — resorts, outfitters, and allied businesses — recreational use did not consume resources.

Indians, fur traders, settlers, prospectors, loggers, fishermen, hunters, canoeists, and managing foresters have all followed BWCA paths. Their footprints vary, their reasons for coming differed, and their impact on the area also varied. But each is part of the BWCA, and as such makes the understanding of human activity essential to an interpretation of this wilderness.

What Is Wilderness?

In seeking wilderness solitude, the rapidly growing number of recreational visitors threatened the very solitude they sought. These visitors endangered the concept of wilderness as a pristine area, free of human influence, to be cherished as the last bastion of the natural, untrammeled world.

The Wilderness Act of 1964 defined wilderness as "an area where man is a visitor who does not remain . . . a community of life untrammeled by man."[4] Wording of this definition was strongly influenced by preservationists, most of whom developed their concept of wilderness in large, remote western areas lacking the history of prior human use so intricately interwoven into the BWCA. Several supervisors of the Superior forest during the 1950s and 1960s believed that the BWCA could not fulfill requirements for such a definition and viewed the area as "an extended sylvan lake and water stage, reconstructed from a populated, logged, and largely burned over, extraordinarily sensitive country."[5] Some wilderness preservationists also believe that the BWCA did not meet the criteria for wilderness and therefore did not belong in the same category as the other, more remote, genuine wilderness areas they championed.[6] They suggest that the management required for the BWCA and the research done there, if classified as wilderness management and research, could be exported to true wilderness areas with dangerous results. The overgeneralization is a two-way street!

Yet the BWCA *is* wilderness. Wilderness is not a geographic entity like a lake, mountain, or river. An area is wilderness if it is perceived as such. The BWCA is wilderness to thousands of people who enter it each year and experience its solitude. Their perception is real, whether earlier human footprints have left permanent changes in the vegetation or not.

Wilderness and People

The 1964 definition of wilderness made people visitors, not part of the wilderness. Since the terms *nature* and *wilderness* are often used synonymously, the definition followed the traditional view of humanity as separate from nature. Nature is perfection that mortal man destroys with his sinful touch. This theme is prevalent in Judeo-Christian writing as, for example, the words of the nineteenth-century hymn writer:

And man's rude work deface no more,
    The paradise of God![7]

This traditional view implies that wilderness can only be maintained if humans get out and stay out. If it can contribute one thing
to the future of our nation's remote forested lands, the unique
mixture of natural and human influences that is the BWCA can
teach us to accept the inevitable reality of human life as part of
nature. It can free us, if not of guilt, at least from the futility of
interpreting wilderness forest succession as proceeding without
human influence.

> The relationship of mankind to nature is, in the light of our traditions,
> ambiguous; we are part of nature, we say, and yet we can control
> nature. We can understand nature, and we can exploit it and devastate
> it, and crush it if we are wont to do so, or if we are simply unthinking
> and careless. We can build a long and enduring future, however, if we
> use knowledge in a rational way – to ask intelligent questions, seek
> answers in intelligent ways, and use intellectual capacities – as
> incorporated into sciences – for the attainment of rational goals and
> a sensible balance of human life with the resources available.[8]

Prior Use and Opposing Concepts

Prior use of the area has complicated the application of wilderness
policies. The BWCA had, after all, been used for timber, wildlife, and
game harvest since the 1700s, had contained land in private ownership with resorts and other establishements, had been traversed by
jeep, truck, motorboat, airplane, and helicopter. Its natural and
human-caused fires had been put out where possible and the pattern
of burning changed in other ways. Trees had been planted, thinned,
shrub growth killed with herbicides, trails and campsites cleared.
Some of these human activities had to be continued and made
acceptable in wilderness legislation; each was hotly contested, and all
contributed to a trammeled condition.

Administration of wilderness policy is further complicated by
the different wilderness concepts still remaining among people in,
near, or interested in the BWCA. Some recreational users still view
the forest with the old fear our ancestors felt. They must have some
semblance of civilization with them when they camp, and few such
visitors venture off a well-marked trail, portage, or campsite; all
cluster their use on lakes close to entry points. Local loggers maintain that, as part of the national forest system, the area should be

available for harvest of timber resources. Sportsmen view the opportunity to fish and hunt, using whatever transportation is most convenient, as their right as citizens. Other recreational users argue that various creature comforts are necessary. Finally, purists believe that wilderness experience is spiritual and that this experience must be primitive and devoid of all human contact. At one time or another, we and most other people who have lived in the BWCA for a long time have, sincerely but ambivalently, adhered to each of these wilderness concepts—sometimes simultaneously—and have thus been accused of speaking with forked tongues.

Wilderness, like beauty, is in the eye of the beholder. The proponents of different wilderness concepts view the BWCA with different eyes. Some are nearsighted, some farsighted, and others have tunnel vision or at least lack three-dimensional sight. Still others are almost blind to the actual realities of the area. Nevertheless, a general agreement has miraculously emerged that wilderness in some condition and size should be maintained and that human activity should not be permitted to destroy it. Much is still left to conceptual interpretation, however.

Like beauty and wilderness, human footprints are also in the eye of the beholder. For some, the crumbling pine stumps left by turn-of-the-century loggers destroy the wilderness impression. For others, lilacs on a former resort site are an insult to their sensibilities. Motorboats, chain saws, and tree-planting crews all appear differently to different people. Elements of the forest mosaic itself are either seen or overlooked, depending on individual interests, knowledge, and ideas.

A Challenged Agency

Wilderness as a philosophic and sociologic paradox became an unprecedented challenge to the Forest Service, the agency responsible for BWCA administration. With each political battle, review, set of public hearings, or change in BWCA status and size, the Forest Service has been charged with devising a new plan of management and implementing it. Many trained staff members have spent countless months developing, revising, and re-revising such plans, only to have them rendered obsolete by a new change of status, congressional act, judicial interpretation, or public appeal for reinterpretation. Rarely has a government agency's task been made more difficult by conflicting influences, opinions, politics, sentiments, and emotions.

Wilderness preservationists often charged the Forest Service with being timber and pulp farmers, too faithful to a strict interpretation of the principles of multiple and prior use. They also accused the agency of adhering too strictly to the prior use exceptions as stated in the 1964 Wilderness Act—those dealing with continued logging and motorboat use. The preservationists had agreed to those exceptions only in order to have the BWCA included in the Wilderness Act, but immediately after passage of the act they brought pressure to have the exceptions ignored in actual management of the area.

In turn, recreationalists, timber producers, and many local residents blamed "the forestry" for restrictions that the agency was legally obligated to enforce. Bitterness, family and neighborhood feuds, vandalism, and blockades resulted. The inability to meet industrial and local needs was not caused by lack of forestry expertise or insensitivity on the part of the agency.

Use of the area had become a political football. Major decisions were forced by politicians very willing to ignore promises made to each side in the conflict. A far cry in complexity from the simple competition among fur traders a century earlier!

Early Steps in Wilderness Preservation

It was a long administrative road from fire protection and managing a remote forest for timber to managing the most heavily used recreational area in the national forest system. That road began nationally with the realization of the excessive damage done by early clearing and timber harvest and the need to preserve forest areas, especially in eastern United States. It prompted Horace Greeley's eloquence in 1850 after his contact with densely populated portions of Europe: "Friends at home! I charge you to spare, preserve, and cherish some portion of your primitive forests; for when these are cut away, I apprehend they will not easily be replaced."[9]

Early wilderness policy is best exemplified by U.S. Congressional Statute at Large 17 of 1872, which set aside Yellowstone Park as the world's first major wilderness preserve. The park was restricted from settlement and all timber, minerals, and natural wonders were to be preserved. Such preservation was for the stated purpose of providing "a public park or pleasuring ground for the benefit and enjoyment of people."[10]

From its inception, then, reservation of wilderness had human enjoyment as its goal. A distinction between preservation of a primitive playground and preservation of wilderness for its own sake evolved slowly. The intrinsic value of wilderness was only realized as wilderness became scarce.

The campaign for wilderness preservation marked a deepening sense of earth stewardship not realized before among a wide spectrum of the population. This new attitude is perhaps the most important contribution of the wilderness movement in the nation. If the old adage "You never miss the water till the well runs dry" holds true, perhaps this sense of stewardship has stimulated many people to ponder what other environmental resources are at the bottom of drying wells.

Although steps were taken to maintain the natural water levels and forests of the border lakes country in the 1920s and early 1930s, it was not until 1938 that portions were classified as Superior Roadless Primitive Area. Names, boundaries, and classification have changed several times since then. However, the identification of the area as something unique from the rest of the Superior National Forest was an important step in preservation of the BWCA's wilderness character.

From Multiple-Use Beginnings

For a time, wilderness regulations were based on an assumption that human enjoyment or recreation could coexist with such material uses as timber harvest and grazing. Although he later changed his views, even John Muir, founder of the wilderness preservationist Sierra Club, said in 1893: "The forests must be and will be not only preserved, but used and . . . like perennial fountains . . . be made to yield a harvest of timber, while at the same time all their far reaching [aesthetic and spiritual] uses may be maintained unimpaired."[11]

These goals for the management of national forest reserves were set in 1905 in Secretary of Agriculture James Wilson's letter to Gifford Pinchot as the national forest system was established. In it, he anticipated the need for conflict resolution:

> In management of each reserve, local questions will be decided on local grounds; the dominant industry will be considered first, but with as little restriction to minor industries as may be possible; sudden changes in industrial conditions will be avoided by *gradual* adjustment *after due notice*; and where conflicting interests must be reconciled, the question will always be decided from the standpoint of the greatest good for the greatest number in the long run. . . . These general principles . . .

can be successfully applied only when the administration of each
reserve is left very largely in the hands of local officers, under the eye
of thoroughly trained and competent inspectors.[12] (emphasis added)

Early in the 1900s, however, most wilderness preservation
leaders believed that logging was not compatible with the wilderness
condition. This belief had its origin in the relentless white pine
harvest that had marched over the northeastern states and had
already begun in the border lakes country. Thus was born the basic
disagreement that ignited the first controversy over management of
the BWCA: the conflict between resource harvest with reforestation
and total forest reservation with no harvest permitted.

The original management of the border lakes forest had incorporated both forest harvest and recreational use. However, like John
Muir, many border lakes recreational users soon despaired of saving
any wilderness quality in the face of logging. They joined other
conservationists who believed that the first, essential step in assuring
wilderness perpetuation was to "save the pieces," to ban cutting and
other resource utilization in largely uncut or "virgin" areas.

Thus, in 1941, the interior no-cut zone comprising 362,000
acres of the Superior National Forest Primitive Area was established.
Since then, Pinchot's "greatest good for the greatest number" within
the border lakes country has given top priority to recreational use —
most recently, strictly primitive recreation. In retrospect, few would
fault the setting aside of the interior no-cut zone. In addition to its
aesthetic and recreational value, the zone is an important natural
laboratory for scientific observation and remains the core of today's
BWCA.

The first step in wilderness preservation, nationwide as well as
in the BWCA, then, was the setting aside of areas from industrial use
and development. This first step, begun in Yellowstone Park, had
evolved in other remote western areas that had little or no previous
commercial use. When the policy was applied to the BWCA, however,
prior use and prior management for timber complicated implementation of wilderness policy and also altered the response of the forest
to the new management plan. As we have seen, past practices had
already altered the resilience of some major tree species and thereby
negated the possibility of natural regeneration of presettlement
forest conditions.

Protection of wilderness as nature manifest for itself rather than
for resource or recreational use was a further change in the wilderness concept. It was to become a nationwide crusade embraced by
wilderness purists with deep spiritual fervor, whether they actually

used a particular wilderness area or not. With the growth of this movement, many Americans were accepting a concept of wilderness that was a complete reversal of the one held by their pioneer ancestors three hundred years ago. This recognition of the intrinsic value of wilderness is noble and worthy of the finest human efforts. However, some of the zealous champions of this concept, like other crusaders as well as like proponents of other wilderness concepts in the past, can be inaccurately informed and easily misdirected. "Nothing does more damage to the prudent management of space and other natural resources than the eloquence of an advocate who does not know what he is talking about."[13]

By the 1960s, wilderness purists from all over the country joined the cause of BWCA preservation. Most organizations and leaders in this powerful movement were from other parts of the state and nation. Many were not familiar with the responses of the BWCA biota to previous human activity in the area.

Management of People in the Wilderness

The publicity generated by various disagreements related to the administration of the BWCA served as a magnet, attracting visitors to the area. The resultant dramatic increase in recreational use brought this realization that human visits must be limited and controlled if the wilderness condition is to be saved. Such limitations on freedom were viewed as the final, albeit necessary, affront to the cherished wilderness experience. Nevertheless, management of people in the wilderness grew from an innovative approach to an absolute necessity. It became a major division of resource management and the subject of university courses, symposia, workshops, and books.

Like the logging restrictions, management of people originated in western areas having little or no prior resource use and light recreational activity. In the BWCA, once more, prior use and management policies complicated the situation and made necessary a phaseout of private holdings, cabins, and resorts and the elimination of mechanized travel in places. It also included camper education, improved campsite maintenance and sanitation, a ban on cans and bottles, and an entry permit system establishing quotas to distribute use and avoid crowding.

As the wilderness preservation movement grew in power and influence, the original Forest Service directive to Pinchot, "local questions will be decided on local grounds," was bypassed. How Pinchot would have chuckled when wilderness movement advocates demanded and got an abrupt ban on wilderness snowmobile use. This ban was contrary to his judicious plan that sudden change in conditions should be avoided by gradual adjustment after due notice, as the Forest Service had intended in its original plans for snowmobile phaseout. The sudden ban caused such a flurry of demonstrations, protests, road blocks, and other defiant gestures that it had to be tempered with more gradual adjustment and compromise. The unpleasantness it created led to more cautious and gradual imposition of future restrictions on wilderness water travel. In dealing with local citizenry, Pinchot, not wilderness preservationists, knew best how to accomplish a goal.

Wilderness Land Management

The idea of manipulating or managing land to maintain or reconstruct wilderness conditions is incongruous with our traditional feelings for the natural state. As such, it has developed slowly and with opposition. The control and direction of forest vegetation as an aspect of wilderness management was first advocated by F. B. Hubachek in 1948 as one of the research goals of the Quetico-Superior Wilderness Research Center (fig. 9.2):

> Wilderness land management is perhaps a new concept. In times past, those areas set aside for recreational use have been managed only in the sense that they have been protected from fire and logging. There is considerable evidence to indicate that this sort of management will not maintain the recreational and aesthetic values peculiar to these areas. Information is sadly lacking as to the over-all ecological effects of various forms of land management including complete protection as well as measures designed to control the environment and direct vegetational changes along desired lines.[14]

In an effort to meet this need for ecological information, the Wilderness Research Center has been conducting biological research in the BWCA for the past thirty-five years, research in which we have been privileged to participate.

In an address to the Sierra Club in 1962, Stephen Spurr concluded that wilderness forests must be managed and modified if they are to be maintained.[15] He also suggested that forest management techniques can reclaim and recreate desperately needed, more extensive wilderness for the future, thus establishing the idea of restorative wilderness management. Most forests included in the eastern wilderness system require this approach because of their prior use.

Figure 9.2. Summer field station of the Quetico-Superior Wilderness Research Center, founded in 1948 by F. B. Hubachek to conduct research related to wilderness vegetation management. Photo by authors.

In 1963, Frank Kaufert championed the cause of wilderness vegetation management and restoration in the keynote address before the Quetico-Superior Institute:

> As a primary goal, we recommend that the biotic association . . . be maintained, or where necessary re-created, as nearly as possible in the condition which prevailed when the area was first visited by white man. Management may at times call for the use of the tractor, chainsaw, rifle, or flame thrower, but the sights and sounds of such activity should be hidden from visitors insofar as possible.[16]

Restorative Wilderness Forest Management

Maintenance or restoration of the primitive condition is not merely a problem of understanding and protecting a resilient forest in its return from clear-cut or natural disturbance. If it were so, management would be relatively simple. Human impact has been too subtle, drastic, and far-reaching for such an approach. It has taken natural momentum out of the forest's return time by destroying seed sources, introducing disease, altering insect outbreak patterns, and changing the whole litter-species allelopathic complex. Yet the BWCA forest succession is still dependent on disturbance for renewal.

In 1963, M. L. Heinselman also supported restorative programs as necessary for wilderness vegetation management.[17] Among these he proposed use of the rifle (hunting), to replace the function of vanished predators in maintaining certain wildlife population balances, and seeding, if cutover lands burned without a seed source. Basically he believed, however, that reintroduction of fire would be all that was needed to restore the original BWCA vegetation.

This suggested reintroduction of fire for restoration of the primeval condition was ill advised because of human alterations in the ecosystem, gaps in the ability to produce and control fire, and lack of information at the time on current fire trends.

Similarly, the advisability of applying other restorative management techniques now, in the light of current knowledge about long-term effects and the forest's damaged resilience, must be approached with caution. However, the possibility of careful, therapeutic timber cutting must be given serious consideration as an alternative to fire for disturbance and forest renewal. Certainly, our ability to control the dangerous aspects of cutting is as great or greater than our ability to control the possible dangers and drawbacks of fire. Both must receive critical, unbiased evaluation; neither would be suitable for every situation needing restoration in the BWCA. In addition, research results thus far indicate that a combination of ecologically sound timber cutting and removal followed by prescribed burning would, especially on some mature jack pine sites, restore a primitive forest better than either fire or timber cutting alone.

We know that past tree planting has reestablished forests in the BWCA that are now indistinguishable from natural forest. For the tall pine species, red and white pine, planting is the only possible means of reestablishment in the BWCA in anything approaching the presettlement condition. Neither species is now being reestablished in significant quantities by natural seeding, and successful artificial seeding would require extensive, prohibitive site preparation. White pine planting must await the development of adequate supplies of strains resistant or tolerant to blister rust, but this development is anticipated.

The reestablishment of these two tall pine species is not merely an aesthetic or nostalgic need. Over one-third of the presettlement forests of the BWCA consisted of stands of these two, the only long-lived species to have occurred in extensive acreages in the BWCA ecosystem. The reestablishment of a good proportion of such long-lived species is essential in the maintenance of maximum biomass and slow, long-term nutrient cycling, in sharp contrast to the rapid biomass turnover and short-term nutrient cycling of the short-lived aspen-birch-spruce-fir complex. Short-term, rapid biomass turnover causes the loss of important nutrients from the system with each rotation, a serious matter in thin, rocky soils. Scattered among the aspen-birch-spruce-fir, stands of tall pine would act as a nutrient bank. As such, their inclusion in the restored wilderness forest is important.

Restorative and maintenance wilderness forest management looms as a future necessity and challenge. The naturally selected and inbred resilience of some BWCA forest species offers great potential for future management. If techniques can be devised to protect and restore resilience to all native species, nature can cooperate with human techniques for restoring something approaching the presettlement forest.

There are many pitfalls to be avoided before restorative management of vegetation can become a practical part of BWCA policy. As in past wilderness plans, there is danger in the application of broad, generalized goals that apply throughout the nation's wilderness system to individual atypical portions of the local area. The presidential directive issued when the Forest Service was instituted in 1905 contained an appeal for individualized, local administration of the forest. This was a very wise caution indeed. The need is also recognized in Europe:

> Planning done in staff headquarters with little attention being paid to flora and fauna of individual sites or to the complicated interplay between the various ecosystem types should be avoided. There is a great risk for impoverishment of the forest landscape associated with the management of large units in such a systematic way, particularly if the foresters' intimate knowledge of his district is replaced by computerized information from systematically-spaced sample plots or aerial photos.[18]

There is a growing trend toward development of management plans by the use of systems analysis and models based on assumptions, use of surveys that do not take long-range changes into consideration, and principles derived from other, largely western ecosystems. However, organisms within different ecosystems are not interchangeable. They differ in responses to environment, usefulness to wildlife, allelopathic potential, and susceptibility to damage from pathogens, pollutants, pesticides, and herbicides. It is, therefore, dangerous to attempt generalizations from one wilderness ecosystem to another without a sound knowledge of species limitations and past history.

At this writing, the fields of systems analysis and model construction in ecology are exciting areas of research, but much work remains to be done before either can become a useful tool in prescribing management of specific wilderness areas. The forester must work with what *is,* not with models that are based on assumptions from the past or from other forests, telling what *could* have been. Only if he does so can he maintain a multispecies forest mosaic with any resemblance to that which would have been here without human use of the area.

There is no past precedent, for example, for an approach to the serious problem of acidification—acid rain. BWCA land plants are already adapted to a somewhat acid soil, and the soil itself acts as a buffer against some of the effects of increasing acidity. The immediate situation for land plants, therefore, does not appear as threatening as for aquatic life. However, the adaptation of many land plants to an acid soil and the buffering action of the soil itself could mask acid accumulation and lull us into complacency until irreversible damage is done. Unless we have long-range, continuing studies monitoring both the terrestrial and aquatic ecosystems, there will be no basis for determining such effects. In addition to possible effects on higher vegetation, the impacts of acidification on soil micropopulation, the all-important lichen-moss stratum or tree seedbed, soil nitrogen relations, and the reserves of buried seed in the forest floor must all be considered. The problem of acid rain demonstrates again that human activity in areas far distant from the BWCA may shape its future.

Looking Back Down the Wilderness Path

Human footprints have led us past lob trees of resource exploitation, harvest, human enjoyment, and recreation to lob trees of conservation and preservation. These all still stand as part of the BWCA forest. The path of the ecology of human use of the BWCA has not ended. At this writing, the border lakes forests are relatively quiet with only primitive recreational use permitted, and that on a limited basis. However, when other resources and other uses of the area arise and their proponents gather in number, power, and influence, controversy among groups with opposing goals will again arise. On the horizon, we hope we can detect more lob trees of wilderness maintenance and restoration, rooted in a sound understanding of BWCA forest ecology and our impact on it and on the realization that human sensitivity and knowledge can at last be at home in the wilderness, no longer a visitor.

There is more at stake here than just the preservation of a million-plus-acre remote forest. Can we learn to walk softly with a healing touch in this delicate, already injured ecosystem, leaving no future scars and possibly removing past ones? Can we live, study, use, and be part of this wilderness in a way that restores both our humanity and the forest? If we can, there is fresh hope that we can do so elsewhere in this universe, and that this goal can replace the vanished frontier as a challenge to brave, wise, creative action.

Appendix

APPENDIX

Common and Scientific Names of Plants

Abbey's hybrid rock fern	*Woodsia abbeae* Butt.
Alder, green	*Alnus crispa* (Ait.) Pursh
Alder, speckled	*Alnus rugosa* (DuRoi) Spreng.
Aspen, trembling or quaking	*Populus tremuloides* Michx.
Asters	*Aster,* various species
Balsam fir	*Abies balsamea* (L.) Mill.
Bishop's cap	*Mitella nuda* L.
Black spruce	*Picea mariana* (Mill.) B.S.P.
Blueberry	*Vaccinium angustifolium* Ait.
Bracken	*Pteridium aquilinum* (L.) Kuhn
Brome grass	*Bromus inermis* Leyss.
Bunchberry	*Cornus canadensis* L.
Burdock	*Arctium minus* (Hill) Bernh.
Buttercup	*Ranunculus,* various species
Chives	*Nothoscordum bivalve* (L.) Britt.
Clintonia	*Clintonia borealis* (Ait.) Raf.
Cloudberry	*Rubus chamaemorus* L.
Columbine	*Aquilegia canadensis* L.
Cranesbill, Bicknell's	*Geranium bicknellii* Britt.
Dandelion	*Taraxacum officinale* Weber var. *palustre* (Sm.) Blytt.
Evening primrose	*Oenothera parviflora* L.
Fescue	*Festuca saximontana* Rydb.
Fireweed	*Epilobium angustifolium* L.
Forget-me-not	*Myosotis scorpioides* L.
Goat's beard	*Tragopogon pratensis* L.

Goldenrod	*Solidago*, various species
Goldthread	*Coptis groenlandica* (Oeder.) Fern.
Goutweed	*Aegopodium podagraria* var. *variegatum* L.
Gypsyweed	*Veronica officinalis* L.
Hazel	*Corylus cornuta* Marsh.
Hepatica	*Hepatica americana* (D.C.) Ker.
Highbush cranberry	*Viburnum trilobum* Marsh.
Hop clover	*Trifolium agrarium* L.
Iris	*Iris pseudacorus* L.
Jack pine	*Pinus banksiana* Lamb.
Juniper, ground (sacred)	*Juniperus communis* L. var. *depressa* Pursh
Lilac	*Syringa vulgaris* L.
Lily-of-the-valley	*Convallaria majalis* L.
Lingonberry	*Vaccinium vitis-idaea* L.
Lupine	*Lupinus perennis* L. var. *occidentalis* S. Wats.
Mints	*Mentha*, various species
Mountain ash	*Pyrus americana* (Marsh.) D.C.
Orange hawkweed	*Hieracium aurantiacum* L.
Orchids	Various genera of the family Orchidaceae (13 genera in the BWCA)
Oxeye daisy	*Chrysanthemum leucanthemum* L. var. *pinnatifidum* Lecoq. & Lamotte
Paper birch	*Betula papyrifera* Marsh.
Peony	*Paeonia lactiflora* Pall.
Plantain	*Plantago major* L.
Poison ivy	*Rhus radicans* L.
Red pine	*Pinus resinosa* Ait.
Redtop	*Agrostis alba* L.
Reindeer moss	*Cladonia rangifera* L. and other species
Rhubarb	*Rheum rhabarbarum* L.
Rush, path-follower	*Juncus tenuis* Willd.
Sheep sorrel	*Rumex acetosella* L.
Shore plantain	*Littorella americana* Fern.
Spring beauty	*Claytonia virginia* L.
Sweet fern	*Comptonia peregrina* (L.) Coult.
Sweet William	*Dianthus barbatus* L.

Tartarian honeysuckle	*Lonicera tatarica* L.
Thistle, Canadian	*Cirsium arvense* (L.) Scop.
Timothy	*Phleum pratense* L.
Violet	*Viola,* several species
Virginia creeper	*Parthenocissus quinquefolia* (L.) Planch.
White cedar	*Thuja occidentalis* L.
White pine	*Pinus strobus* L.
White spruce	*Picea glauca* (Moench.) Voss.
Wild plum	*Prunus americana* Marsh.
Willow	*Salix,* several species
Wood anemone	*Anemone quinquefolia* L.
Yarrow	*Achillea lanulosa* Nutt.
Yellow dock	*Rumex crispus* L.

Notes

Notes

Chapter 1. Red Pine Lob Tree

1. James Wilson, Secretary of Agriculture, in a 1905 letter to Gifford Pinchot, first chief of the Forest Service, when that agency was established by President Theodore Roosevelt.

2. J. C. Hendee, G. H. Stankey, and R. C. Lucas, *Wilderness management*, U.S. Forest Service Miscellaneous Publication 1365, 1978, p. 117.

3. R. Nash, *Wilderness and the American mind*, rev. ed. (New Haven: Yale University Press, 1973), pp. 8-22.

4. Cited in J. A. Zivnuska, The managed wilderness, *American Forests* 79(8):16 (1973).

5. H. D. Thoreau, *The Maine woods*, 1864. Reprinted in *The writings of Henry David Thoreau*, vol. 3 (New York: First AMS Press, 1968), p. 71.

6. H. D. Thoreau, Walking, *Atlantic Monthly*, June 1862. Reprinted in *The Writings of Henry David Thoreau*, vol. 5, p. 224.

7. C. Rupp, personal communication, 1980.

8. Cited from the masthead of *Synergist*, the official publication of the Superior National Forest.

Chapter 2. Sacred Juniper Lob Tree

1. J. C. Hendee, G. H. Stankey, and R. C. Lucas, *Wilderness management*, U.S. Forest Service Miscellaneous Publication 1365, 1978, p. 237.

2. W. Upham, *Catalogue of the flora of Minnesota*, Geological and Natural History Survey of Minnesota, Part 4, Annual Report of Progress, 1883 (Minneapolis: Johnson, Smith, and Harrison Co., 1884), p. 15.

3. C. E. Ahlgren and I. F. Ahlgren, Some effects of different forest litters on seed germination and growth, *Canadian Journal of Forest Research* 11:710-14 (1981).

4. S. Walshe, *Plants of the Quetico and the Canadian Shield* (Toronto: University of Toronto Press, 1980), p. 126.

5. T. Morley, Rare or endangered plants of Minnesota with the counties in which they are found, Department of Botany, University of Minnesota, St. Paul, 1972.

6. C. E. Ahlgren, Phenological observations of nineteen native tree species in north-eastern Minnesota, *Ecology* 38:622-28 (1957).

7. Upham, *Flora of Minnesota,* 1884.

8. O. Lakela, *Flora of northeastern Minnesota* (Minneapolis: University of Minnesota Press, 1965).

9. I. F. Ahlgren and C. E. Ahlgren, *Revised checklist of ferns and flowering plants of the Quetico-Superior Wilderness Research Center,* Q-SWRC Technical Note 4, 1962.

10. Walshe, *Plants of the Quetico,* 1980.

11. Unpublished checklists of plants of several Minnesota counties prepared by J. W. Moore, G. B. Ownbey, and J. P. Eman are on file with the Department of Botany, University of Minnesota, St. Paul.

Chapter 3. Jack Pine Lob Tree

1. J. Fitzwater, tape-recorded interview by R. Naddy, information specialist, Superior National Forest, in *Historical sketches of the Quetico-Superior,* vol. 12, compiled by J. W. White, Superior National Forest, U.S. Forest Service, 1970, p. 3.

2. C. O. Rosendahl, *Trees and shrubs of the Upper Midwest* (Minneapolis: University of Minnesota Press, 1955), p. 52.

3. A. M. Swaim, A history of fire and vegetation in northeastern Minnesota as recorded in lake sediments, *Quaternary Research* 3:383-96 (1973).

4. A. J. Craig, Pollen influx to laminated sediments: A pollen diagram for northeastern Minnesota, *Ecology* 53:46-57 (1972).

5. M. L. Heinselman, Fire in the virgin forests of the Boundary Waters Canoe Area, Minnesota, *Quaternary Research* 3:329-82 (1973).

6. G. T. Woods and R. J. Day, *Fire ecology study, Quetico Provincial Park,* Reports 1-7, Atikokan District, North Central Region, Ontario Ministry of Natural Resources, 1976-1977.

7. H. Hansen, College of Forestry, University of Minnesota, personal interview.

8. W. J. Emerson, correspondence, 1981-1982; D. A. Haines and E. L. Kuehnast, When the Midwest burned, *Weatherwise* 23:112-19, (1970).

9. G. L. Nute, *The Voyageur's Highway,* (St. Paul: Minnesota Historical Society, 1951), p. 75.

10. Heinselman, Fire in the virgin forests, 1973.

11. Woods and Day, *Fire ecology study,* 1976-1977.

12. F. J. Marschner, Original forests of Minnesota (map), 1930, published by North Central Forest Experiment Station, U.S. Forest Service, St. Paul, Minnesota, 1974.

13. A. R. Taylor, Lightning effects on the forest complex, *Proceedings Tall Timbers Fire Ecology Conference* 9:127-50 (1969).

14. D. E. Olson, Physics Department, University of Minnesota, Duluth, personal interview, 1981.

15. P. Kourtz, *Lightning behavior and lightning fires in Canadian Forests,* Forestry Branch, Department of Forestry and Rural Development, Canada, Departmental Publication 1179, 1967.

16. G. D. Freier, Department of Physics, University of Minnesota, Minneapolis, correspondence, 1981; D. M. Fuquay, R. G. Baughman, and D. J. Latham, *A model for predicting lightning fire ignition in wildland fuels,* U.S. Forest Service Research Paper INT-217, 1979.

17. W. J. Emerson, correspondence, 9 November 1981.

18. L. F. Ohmann and D. F. Grigal, Contrasting vegetation responses following two forest fires in northeastern Minnesota, *American Midland Naturalist* 106:54-64 (1981).

19. C. E. Ahlgren, *Eighteen years of weather in the Boundary Waters Canoe Area*, Agricultural Experiment Station, University of Minnesota, Miscellaneous Report 89, 1969; Annual Fire Log, U.S. Forest Service, unpublished.

20. W. J. Emerson, correspondence, 27 October 1981.

21. Heinselman, Fire in the virgin forests, 1973.

22. M. J. Schroeder and C. C. Buck, *Fire weather . . . a guide for the application of meteorological information to forest fire control measures,* U.S. Forest Service Agricultural Handbook 360, pp. 197-203, 1970; M. J. Schroeder, *The Hudson Bay High and spring fire season in the Lake States,* U.S. Forest Service Fire Control Series 110:1-8 (1950).

23. W. J. Emerson, correspondence, 27 October 1981.

24. Heinselman, Fire in the virgin forests, 1973, p. 330.

25. E. V. Komarek, Sr., The nature of lightning fires, *Proceedings Tall Timbers Fire Ecology Conference* 7:5-41 (1967).

26. W. J. Emerson, correspondence, 9 December 1981.

27. C. E. Ahlgren, Regeneration of red and white pine following wildfire and logging in northeastern Minnesota, *Journal of Forestry* 74:135-40 (1979).

28. C. E. Ahlgren, *Some effects of prescribed burning on jack pine reproduction in northeastern Minnesota,* Agricultural Experiment Station, University of Minnesota, Forestry Series 5, Miscellaneous Report 94, 1970.

29. C. E. Ahlgren, Effects of fires on temperate forests: North central states, in *Fire and ecosystems,* edited by T. T. Kozlowski and C. E. Ahlgren (New York: Academic Press, 1974), pp. 195-219.

30. J. O. Nordin and D. F. Grigal, Vegetation, site, and fire relationships within the area of the Little Sioux Fire, northeastern Minnesota, *Canadian Journal of Forestry* 6:78-85 (1976).

31. T. J. Carleton and P. F. Maycock, Vegetation of the boreal forests south of James Bay, *Ecology* 61:1199-1212 (1980).

32. C. E. Ahlgren, Some effects of fire on reproduction and growth of vegetation in northeastern Minnesota *Ecology* 41:431-45 (1960).

33. C. E. Van Wagner, Fire and red pine, *Proceedings Tall Timbers Fire Ecology Conference* 10:211-19 (1970).

34. Woods and Day, *Fire ecology study,* 1976-1977.

Chapter 4. Big Cedar Lob Tree

1. J. E. Potzger, History of the forests in the Quetico-Superior country from fossil pollen studies, *Journal of Forestry* 51:560-65 (1953).

2. A. M. Swaim, A history of fire and vegetation in northeastern Minnesota as recorded in lake sediments, *Quaternary Research* 3:383-96 (1973).

3. A. J. Craig, Pollen influx to laminated sediments: A pollen diagram from northeastern Minnesota, *Ecology* 53:46-57 (1972); D. E. Amundson and H. E. Wright, Jr., Forest changes in Minnesota at the end of the Pleistocene, *Ecological Monographs* 49:1-16 (1979).

4. J. W. White, *Forest fires in the Quetico-Superior country in the 17th and 18th centuries and before: Historical Notes of the Superior National Forest,* U.S. Forest Service, 1965.

5. D. D. Owen, *Report of a geological survey of Wisconsin, Iowa, and Minnesota* (Philadelphia: Lippincott, Grambo & Co., 1892).

6. P. Kane, *Wanderings of an artist among the Indians of North America from Canada to Van Couver's Island and Oregon through Hudson's Bay Company territory and back again,* 1858 (reprinted Rutland, Vermont: Charles E. Tuttle, 1968), p. 41.

7. J. B. Bigsby, *The shoe and canoe, or pictures of travel in the Canadas* (London: Chapman and Hall, 1850).

8. J. W. White, *General land office surveys, 1858-1907: Historical sketches of the Quetico-Superior*, vol. 9, Superior National Forest, U.S. Forest Service, 1970, pp. 5-8.

9. White, *General land office surveys*, 1970, p. 6.

10. J. T. Curtis, *The vegetation of Wisconsin* (Madison: University of Wisconsin Press, 1959), pp. 457-72.

11. F. J. Marschner, The original vegetation of Minnesota (map), 1929, published with interpretation by M. L. Heinselman by the North Central Forest Experiment Station, U.S. Forest Service, St. Paul, Minnesota, 1974.

Chapter 5. Paper Birch Lob Tree

1. G. L. Nute, *Caesars of the wilderness* (New York: Appleton-Century Company, 1943).

2. M. E. Cambell, *The Nor'westers, the fight for the fur trade* (Toronto: Macmillan Co., 1954).

3. J. W. White, *Fish and wildlife along the Voyageurs' Highway: Historical sketches of the Quetico-Superior*, Superior National Forest, U.S. Forest Service, 1965.

4. G. L. Nute, *The Voyageur's Highway* (St. Paul: Minnesota Historical Society, 1951), p. 66.

5. M. Stenlund, correspondence, July 1981. See also W. H. Longley and J. B. Moyle, *The beaver in Minnesota*, Minnesota Department of Conservation Technical Bulletin 6, 1963.

6. Nute, *Voyageur's Highway*, 1951, p. 61.

7. M. L. Heinselman, Fire in the virgin forests of the Boundary Waters Canoe Area, Minnesota, *Quaternary Research* 3:329-82 (1973).

8. White, *Fish and wildlife*, 1965, p. 7.

9. J. J. Bigsby, The shoe and canoe, or pictures of travels in the Canadas (London: Chapman and Hall 1850), p. 245.

10. J. W. White, *General Land Office Surveys, 1858-1907: Historical Sketches of the Quetico-Superior*, vol. 9, Superior National Forest, U.S. Forest Service, 1970, p. 17.

11. C. E. Ahlgren, Some effects of fire on reproduction and growth of vegetation in northeastern Minnesota, *Ecology* 41:431-45 (1960).

Chapter 6. White Pine Lob Tree

1. A. M. Larson, History of the white pine industry in Minnesota (Minneapolis: University of Minnesota Press, 1949).

2. J. W. Trygg, *Swallow and Hopkins Lumber Company of Winton, Minnesota, A brief resume of events that took place*, and G. H. Good, *Pioneer logging*, privately distributed by Ely-Winton Historical Society, 1966.

3. Mr. and Mrs. Carl Gawboy, personal communication, 1981.

4. M. L. Heinselman, Fire in the virgin forests of the Boundary Waters Canoe Area, Minnesota, *Quaternary Research* 3:329-82 (1973).

5. Personal communications with W. J. Trygg, 1960; F. H. Kaufert, 1981; and J. Kernick, 1981.

6. *Duluth News-Tribune*, 18 May 1909, p. 4.

7. Heinselman, Fire in the virgin forests, 1973.

8. S. Walshe, correspondence, 1981.

9. C. E. Van Wagner, Fire and red pine, *Proceedings Tall Timbers Fire Ecology Conference* 10:211-19 (1970).

10. D. P. Fowler and R. W. Morris, Genetic diversity in red pine: Evidence for low genetic heterozygosity, *Canadian Journal of Forest Research* 7:343-47 (1977).

11. D. P. Fowler and D. T. Lester, *The genetics of red pine,* U.S. Forest Service Research Paper WO-8, 1970.

12. C. E. Ahlgren and I. F. Ahlgren, Some effects of different forest litters on seed germination and growth, *Canadian Journal of Forest Research* 11:710-14 (1981).

13. C. E. Ahlgren, Some effects of fire on reproduction and growth of vegetation in northeastern Minnesota, *Ecology* 41:431-45 (1960).

14. H. L. Hansen, E. V. Bakuzis, and V. Kurmis, *Effects of fire, logging, and herbicides on the forest ecosystem,* Department of Forest Resources, College of Forestry, University of Minnesota, Research and Extension Report, 1979-1980.

15. L. B. Ritter, Sr., unpublished papers, 1981.

16. R. G. Pierce, Early discoveries of white pine blister rust in the United States, *Phytopathology* 7:224-25 (1917).

17. E. E. Honey, *East Pike Lake pine infection study plot, 1939-1940 report,* U.S. Forest Service, Superior National Forest, 1940.

18. D. B. King, *Incidence of blister rust infection in the Lake States,* U.S. Forest Service, Lake States Forest Experiment Station Paper 64, 1958.

19. L. B. Ritter, Sr., White pine for the urban landscape, *The Minnesota Horticulturist* 110:100-101 (1982).

20. I. F. Ahlgren, New hope for eastern white pine, *American Forests* 85:46-49, 58-62 (1979).

21. N. T. Mirov, *The genus Pinus* (New York: The Ronald Press, 1967), p. 101.

22. J. V. Thirgood, *Man and the Mediterranean forest* (New York: Academic Press, 1981), pp. 67-80.

23. F. J. Marschner, Original forests of Minnesota (map), 1930, published by North Central Forest Experiment Station, U.S. Forest Service, St. Paul, Minnesota, 1974.

24. D. Kee, Selected letters, in *Historical sketches of the Quetico-Superior,* vol. 7, compiled by J. W. White, Superior National Forest, U.S. Forest Service, 1969, p. 17.

25. C. E. Ahlgren and I. F. Ahlgren, The human impact on northern forest ecosystems, in *The Great Lakes forest: an environmental and social history,* ed. S. Flader (Minneapolis: University of Minnesota Press, 1983), 33-51.

26. M. L. Heinselman, The extent of natural conversion to other species in the Lake States aspen-birch type, *Journal of Forestry* 52:737-38 (1954).

27. Sather, N. Vegetation within a portion of the Copper-Nickel study region, *Journal of the Minnesota Academy of Science* 46(2):19-21 (1980).

28. Ahlgren and Ahlgren, Some effects of different forest litters, 1981.

29. R. T. Franklin, Insect influences on forest canopy, in *An analysis of temperate forest ecosystems,* edited by D. E. Reichle (New York: Springer-Verlag, 1970), pp. 86-99.

30. G. D. Van Raalte, Do I have a budworm-susceptible forest? *The Forestry Chronicle* 48:190-92 (1972).

31. Ahlgren and Ahlgren, Some effects of different forest litters, 1981.

32. J. J. de Gryse, Enemies of the forest—Man or insects, *Proceedings Royal Canadian Institute,* series 3A, vol. 9, session 1943-1944, pp. 52-62.

33. J. R. S. Blais, Spruce budworm outbreaks in the past three centuries, *Forest Science* 11:130-38 (1965); D. P. Duncan and A. C. Hodson, Influence of forest tent caterpillar upon the aspen forests of Minnesota, *Forest Science* 4:71-93, (1958).

34. Ahlgren and Ahlgren, Some effects of different forest litters, 1981.

Chapter 7. White Spruce Lob Tree

1. D. Kee, Communication, in *Historical sketches of the Quetico-Superior,* vol. 7, compiled by J. W. White, Superior National Forest, U.S. Forest Service, 1969, p. 17.
2. R. St. Amant, interview, 1981.
3. D. H. Alban, D. A. Perala, and B. E. Schlaegel, Biomass and nutrient distribution in aspen, pine, and spruce stands on the same soil type in Minnesota, *Canadian Journal of Forest Research* 8:290-99 (1978); J. P. Kimmins, Evaluation of the consequences for future tree productivity of the loss of nutrients in whole tree harvesting, *Forest Ecology and Management* 1:169-83 (1977).
4. T. R. Crow and R. W. Blank, Distribution of biomass and productivity for several northern woody species, U.S. Forest Service, North Central Forest Experiment Station Research Note NC-239, 1978.
5. R. Hagman, interview, 1982.
6. Interviews with R. Hagman, J. Kernick, and W. "Pappy" Wright, 1981-1982.
7. S. Rommel, correspondence, 1982.
8. N. T. Mirov, *The Genus Pinus,* (New York: Ronald Press, 1967), p. 451.
9. J. App, interview, 1981.
10. R. St. Amant, interview, 1981.
11. J. Kernick, interview, 1982.

Chapter 8. Balsam Fir Lob Tree

1. C. E. Ahlgren and I. F. Ahlgren, Some effects of different forest litters on seed germination and growth, *Canadian Journal of Forest Research* 11:710-14 (1981).
2. E. V. Bakuzis and H. L. Hansen, *Balsam fir,* (Minneapolis: University of Minnesota Press, 1965), p. 31.
3. M. Fries, Pollen profiles of late Pleistocene and recent sediments from Weber Lake, Minnesota, *Ecology* 43:295-308 (1962).
4. Cited in J. A. Larsen, *The boreal ecosystem* (New York: Academic Press, 1980), p. 174.
5. W. H. Richardson, A honeymoon in a birch bark canoe, 1897, in *Historical sketches of the Quetico-Superior,* vol. 11, compiled by J. W. White, Superior National Forest, U.S. Forest Service, 1974, p. 34.
6. J. W. Trygg, *Pioneer logging. A brief resume of events that took place: Swallow & Hopkins Lumber Company and the Fall-Basswood Lake logging railroad,* unpublished paper from the historical collection of J. W. Trygg, 1952, p. 7.
7. F. A. Waugh, *Recreational uses on the national forests,* report to the U.S. Forest Service, Washington, D.C., 1918.
8. A. H. Carhart, *Preliminary prospectus: An outline plan for the recreational development of the Superior National Forest,* unpublished report in the historical files of the Superior National Forest, U.S. Forest Service, 1921.
9. R. N. Searle, *Saving Quetico-Superior, a land set apart* (Minneapolis: Minnesota Historical Society Press, 1977); J. C. Hendee, G. H. Stankey, and R. C. Lucas, *Wilderness management,* U.S. Forest Service Miscellaneous Publication 1365, 1978.
10. A. H. Carhart, A forest home for everyone, *Good Housekeeping Magazine,* June 1920.
11. J. W. White, *Historical sketches of the Quetico-Superior,* vol. 6, p. 2, Superior National Forest, U.S. Forest Service, 1968.
12. All figures used in discussions of visitor numbers, length of stay, mode of travel, and purpose of trip are taken from Recreational Use Statistics, Forms 446, and Annual Statistical Report, BWCA, Superior National Forest, U.S. Forest Service.

13. H. E. Andersen, correspondence, 1981.

14. R. Nash, *Wilderness and the American mind,* rev. ed. (New Haven: Yale University Press, 1973), pp. 96-236.

15. L. H. Graber, *Wilderness as a sacred place,* Association of American Geographers, Washington, D.C., Monograph Series, 1976, p. 111.

16. J. A. Zivnuska, The management of wilderness, *American Forests* 79:16-19, 41 (1973), p. 16.

17. U.S. Forest Service figures; see note 12.

18. A. Leopold, *A Sand County almanac* (New York: Oxford University Press, 1949), p. 203-4.

19. Nash, *Wilderness and the American mind,* 1973.

20. Superior National Forest campsite map and report, spring 1982.

21. S. S. Frissell, Jr., and D. P. Duncan, Campsite preference and deterioration in the Quetico-Superior canoe country, *Journal of Forestry* 63:256-60 (1965).

22. L. C. Merriam, Jr., and C. K. Smith, Visitor impact on newly developed campsites in the BWCA, *Journal of Forestry* 72: 627-30 (1974).

23. R. C. Lucas, *Recreational use of the Quetico-Superior area,* U.S. Forest Service Research Paper LS-8, 1964, p. 1.

Chapter 9. Aspen Lob Tree

1. J. W. White, *Historical sketches of the Quetico-Superior,* vol. 12, Superior National Forest, U.S. Forest Service, 1969, p. 65.

2. J. W. White, *Historical sketches of the Quetico-Superior,* vol. 10, Superior National Forest, U.S. Forest Service, 1968, p. 6.

3. M. L. Heinselman, The extent of natural conversion to other species in the Lake States aspen-birch type, *Journal of Forestry* 52:737-38 (1954).

4. Public Law 88-577, Sec. 2-c, U.S. Congress, 1964.

5. U.S. Forest Service, *Superior National Forest Handbook,* 1964, p. 9.

6. R. Costley, An enduring resource, *American Forests* 78(6):8-11, 54-56 (1972).

7. Charles Kingsley, 1871; hymn 315, *The Hymnbook,* Presbyterian Church in the U.S.A.

8. J. A. Larsen, *The boreal ecosystem* (New York: Academic Press, 1980), p. 442.

9. Cited in R. Nash, *Wilderness and the American mind,* rev. ed. (New Haven: Yale University Press, 1973), p. 96.

10. U.S. Statutes at Large 17, 1872.

11. A plan to save the forests, *Century* 49:631 (1895).

12. Cited in J. C. Hendee, G. H. Stankey, and R. C. Lucas, *Wilderness management,* U.S. Forest Service Miscellaneous Publication 1365, 1978.

13. B. F. Lamb, Elements of wilderness management, *Journal of Forestry* 71:589 (1973).

14. F. B. Hubachek, Objectives of the Quetico-Superior Wilderness Research Center, 1948, p. 3; privately distributed.

15. S. H. Spurr, Wilderness management, in *Tomorrow's wilderness,* edited by F. Leydet (San Francisco: Sierra Club, 1963).

16. F. H. Kaufert, Keynote address, Quetico-Superior Institute, Minneapolis, 1964.

17. M. L. Heinselman, Fire in the virgin forests of the Boundary Waters Canoe area, Minnesota, *Quaternary Research* 3:329-82 (1973).

18. O. O. Tamm, *Modern mechanized forest management—Does it violate ecological principles?* (English abstract) Swedish University of Agricultural Science, Department of Ecology and Environmental Research Report 3, Uppsala, p. 11.

Index

Index

CLIFFORD AHLGREN and ISABEL AHLGREN worked for many years at the Wilderness Research Foundation in Ely, Minnesota, and were also research associates in the College of Forestry at the University of Minnesota. They lived and worked in the Boundary Waters Canoe Area and have written extensively on forest succession and the ecology of the boreal zone in northern Minnesota. They now live in Sun City, Arizona.